FIRE COMMAND

Alan V. Brunacini

National Fire Protection Association
Batterymarch Park, Quincy, MA 02269
NFPA®

Produced by:

 PRODUCTIONS

PO BOX 775 • COLLEGE PARK, MARYLAND 20740

Executive Producer: Jim Yvorra
Art Director/Cover Design: Don Sellers, AMI
Educational Technologist: J. David Bergeron
Production Coordinator/Text Design: LaWan Sellers
Editor: Janis Oppelt
Art Assistant: Leslie Twohig

Typesetter: Key Printing, Landover, MD
Typeface: Rockwell Light

NFPA No. FSP-70

ISBN 0-87765-284-8

Library of Congress Card No. 84-61415

Printed in the USA.

95

DEDICATION

To: Battalion Chief Dario Travaini,
the first and best FGC

ABOUT THE AUTHOR

Alan V. Brunacini has been a member of the Phoenix Fire Department since 1958 and was appointed Fire Chief in 1978. He is a member of the NFPA Board of Directors and is Chairman of the NFPA Fire Service Occupational Safety and Health Committee. He is a graduate of Oklahoma State University's School of Fire Protection and holds Bachelor of Science and Master of Public Administration degrees from Arizona State University. Chief Brunacini has been a student of firefighting operations throughout his career. He lives in Phoenix with his family, a good natured mastiff, a grumpy Burmese and a 1952 Mack Pumper.

TABLE OF CONTENTS

INTRODUCTION
MEDIOCRE DAYS AND BAD NIGHTS

This text presents a great deal of charts, graphs, models, lists, and check sheets which outline and describe the Fireground Command System. The material represents an attempt to assemble the concepts and practices that will help a Fireground Commander (FGC) do his job by applying fundamental management principles to firefighting operations.

These principles describe the theory that must be applied while firefighting practices describe the work that must be done. The manager in us wants to routinize, control, and think; the firefighter in us wants to act, react, and feel. The effective operation of the system requires a mixture of both.

The FGC is in the position of applying the command system to situations that have almost limitless combinations of factors which all come together on the fireground. This practical setting becomes the real strategic and tactical world and always presents a challenge to him. He quickly discovers the natural difference between what is presented in his educational process and what he actually encounters. His challenge is to understand, and then apply these lessons on the fireground.

Normal fireground confusion is often complicated by the following realities:

- Fires grow and move in highly dynamic ways
- The arrangement and construction of structures create a variety of barriers and problems
- Human beings tend to exhibit screwy behavior under fire conditions.

These factors have a natural resistance to being lined up into a management system. Their uncooperative reaction creates a difficult situation for the FGC that cannot be described in a textbook. The entire command system is an attempt to somehow intellectualize a fast-moving and violent event. These major components come together in a way that is usually difficult (sometimes impossible) to fit into neat, well-arranged cubbyholes which are easy to understand and control.

The attempt to represent what firefighting and fire command "feel like" becomes a most difficult task. It is roughly equivalent to constructing a written description of what jamoca almond fudge ice cream tastes like—it loses something in the translation.

There are certain times when it seems like nothing goes right for the FGC. The fire doesn't react according to plan, the building was built to burn, and everyone on the fireground acts like there is a full moon. A realistic command system must anticipate all the things that can, and routinely do, go wrong on the fireground. These are the times when the FGC must strongly support and protect his troops and also work very hard to achieve his tactical objectives. The ability to make the system work on these mediocre days and bad nights truly defines the FGC.

A Fireground Commander who is successful most of the time on most of his fires will:

1. Use the standard elements present at every fire to establish and maintain command.

Fireground factors provide an inventory of the standard conditions present on every fire scene, although their combination is clearly different from fire to fire. The factor inventory serves as a check list the FGC uses for evaluating, decision making, and creating effective action. This information management becomes the basis for the command process. Fires are always strange and surprising for the FGC who regards every fire as a completely unique event. Standard factors give the FGC the capability to continually refine his management ability by depositing his experiences into an information bank. Starting operations from a common beginning point and applying the collective wisdom gained from past fires is a big asset to the FGC.

2. Be responsive to the special characteristics of each situation.

Athough standard factors provide the building blocks, the FGC must always develop an attack plan that is customized to fit his current needs. His ability to beat the fire depends on his ability to develop decisions based on fast, accurate evaluation and then to forecast and manage information to keep the plan current. This ability is the result of a balance between study, experience and reflection. Fire command is an exercise in mixing and matching the standard factors of conditions and action to current needs. This challenge will always make the job of the FGC more exciting than being a bank teller (unless there is a holdup in progress).

3. Apply the command system to every fire situation.

In spite of all the normal fireground confusion, the smart money will always bet on the FGC who consistently uses standard command practices to framework his efforts. These procedures create a system where everyone on the team knows what to expect, where they will fit in, and what the rest of the team will do. The FGC leads and supports the team while continually reinforcing the dependability of the attack process.

The times that really require everyone to be in the right place doing the right thing are the bad nights when fire victims are jumping out of the windows and the parapets are falling. Tough times like these make working the command system difficult, yet they are when the system is needed the most. The FGC now has a security blanket of standard functions that will get him into and, eventually, out of rescue and fire control operations.

Command Reflection

Applying lessons learned to future fires is a crucial FGC performance yardstick. Thinking about the command process produces a perspective that translates preparation and experience into performance. Everything on the fireground involves action—both good and bad. Sometimes it occurs so fast and is so intense that the FGC can only sort out and summarize the action. These mental gymnastics become a personal technique to integrate thinking and fireground action (action without thinking = ineptness; thinking without action = impotency). This combination of experience and reflection causes the FGC to connect the beginning, middle, and conclusion of fire scenarios. There is simply no substitute for this do it, think about it, forecast it, do it again, think about it . . . approach.

Managing past fires can never produce improved command performance unless they are completely processed by the FGC. This reflection also gives him the stability to comprehend what happened and to realize that the sun always comes up at the end of a bad night.

ACKNOWLEDGMENTS

It would take well over 65 pages to thank all the people who have helped produce this book. The effort has taken 25 years and is the result of being active with virtually every firefighter within a fire department that has gone from 2,500 alarms in 1958 to about 75,000 in 1984. This material has emerged from instructing community college tactics classes and presenting the FGC seminar around the country. These sessions constantly taught me far more than the students.

I extend my special thanks to all members of the Phoenix Fire Department. They have survived all the changes involved in moving from a basically unstructured life on the fireground to a very active program using standard operating procedures. Phoenix firefighters perform in a professional, aggressive, and positive manner. Interacting with them has been a great joy in my life. I would be derelict if I did not mention the members of the 'B' shift. They have an unusually good time making the system work, and their overall approach to life gives them the ability to entertain themselves, even in a stuck elevator.

The individual who is most responsible for getting me through life is my wife, Rita. If it were not for her influence, I would have long ago joined a carnival and would no doubt today be hustling the little cutter gadgets that make radishes look like tulips. Her intelligence, humor, coolheadness, and philosophical approach would have made her an excellent FGC.

My two 'B' shift (naturally) firefighter sons, Nick and John, have given me the chance to rejoin the fire department from the practical perspective available only at the end of the nozzle. My daughter, Candi, has supplied me with an endless array of off-color podium jokes and stories. I thank them for their patience when I yelled "turn down that damn TV—I'm trying to write a textbook that will give mankind hope."

My secret weapon in this project has been J. Gordon Routley, the man who has taken the first and last cut at my hallucinations for many years. His efforts range from minor tune-ups to major overhauls, and he has held up to the task well. He is the most literate and technically competent actor in today's fire service. I thank him for his editing, assistance, and Canadian dry humor. His observations are always on the mark and always critical enough to keep my ego in check.

My seminar partners, Assistant Chief Chuck Kime and Division Chief Bruce Varner, have made major contributions and refinements to the Fireground Commander material over the years. We have had the FGC seminar on the road since 1974 and have suffered through American Airlines chicken, rental cars with miniature trunks, and Holiday Inns with screens too small to see when sitting in the back of the room. They have humored me while I scribbled notes in hallways and poorly lighted restaurants all over America. They are good people and an integral part of the Phoenix command system.

A very special group of friends acted as content reviewers for the many drafts of the text. In alphabetical order they are:

Norm Angelo, Chief
Kent Fire Department
Kent, Washington

Jack Bennett, Assistant Chief
Los Angeles City Fire Dept.
Los Angeles, California

Robert Bingham,
   Battalion Fire Chief
District of Columbia Fire Dept.
Washington, D.C.

Dennis Dewar, Fire Marshall
State of Florida
Tallahassee, Florida

M.H. "Jim" Estepp, Chief
Prince George's County
   Fire Department
Upper Marlboro, Maryland

A.J. Evans, Deputy Director
Fire Academy
Justice Institute of
   British Columbia
Vancouver, B.C.

Jackson Gerhart, Firefighter
District of Columbia Fire Dept.
Washington, D.C.

Hank Howard, Chief
Benecia Fire Department
Benecia, California

Michael Hildebrand,
   Coordinator/
Safety and Fire Protection
American Petroleum Institute
Washington, D.C.

William Killen, Chief
Public Safety Division
Washington National Airport
Washington, D.C.

Ed McCormack,
   Executive Director
International Society of
   Fire Service Instructors
Ashland, Massachusetts

Max H. McRae, District Chief
Houston Fire Department
Houston, Texas

Richard Schramm, Director
Bucks County Fire Academy
Doylestown, Pennsylvania

Allan Sullivan, Instructor
United States Park Police
Federal Law Enforcement
   Training Center
Glynco, Georgia

G. Crawford Wiestling,
   Investigator
Fetterly & Purdy
Minneapolis, Minnesota

Dan Young, Chief
Dale City Volunteer Fire Dept.
Dale City, Virginia

Several of these experienced reviewers deserve an extra salute for adding substantially to the organization or content: Jack Bennett, Mike Hildebrand, Max McRae, and Rich Schramm.

Regardless of how long and hard you write, there is no other format, including magazine articles, that exactly matches book form. The publishing professionals who have massaged and assembled the command material and made it look and act like a book are the YBS Productions team—producer Jim Yvorra, art director Don Sellers, and educational technologist J. David Bergeron. LaWan Sellers was also a gigantic addition to the entire development and printing effort. Their special combination of talent and experience have proven again the need to have the right people working in the right sectors. The process of giving birth is always painful (and rewarding), and YBS has been a highly capable literary midwife.

Like every FGC, I assume responsibility for the overall outcome, especially any mistakes in the material.

AVB

HOW TO USE THIS BOOK

FIRE COMMAND is written to present an easy to understand, comprehensive outline for effective, standard command operations at the scene of any emergency. Your course will follow local guidelines, but it will be similar to command level courses offered throughout the country.

This text can point out the facts that apply to all fires and can tell you how to collect the facts you need for each specific fire. It also can help you to develop the discipline needed for Command by showing you the standard approaches to common problems. Even though there is some practice involved with any textbook and workbook, the practice required to make you an effective Fireground Commander comes from case studies, simulations, the fireground, and a lot of late night thinking. Getting this practice is your job.

The basic tactics of fire fighting are also presented. Each is related to the strategic functions of Command. As you read this text, look to see how strategy defines tactics.

Firefighters work at the task level. They have assigned work to do and a specific time in which to complete it. A Fireground Commander does not have to know how to do every job on the fireground, but he should know what jobs are being done and have a general understanding of how they are done and how long they will take.

One of your tasks as a student is to recognize this text as a framework upon which to build your department's operations (strategies, tactics, and tasks). The final authority in your training is not this text, it is your instructor or Training Officer. Each department continually modifies old procedures and adopts new ones. There is no way we can cover all departmental procedures or changes for the entire country. In cases where this text and your instructor take different approaches, follow your instructor.

OBJECTIVES

Each chapter and chapter section begins with a list of specific objectives. The objectives list helps you to flag what you already know and indicates what to emphasize in your first reading. Once you have read the chapter, the objectives serve as a personal test to determine if you have understood what was presented. Note that page numbers appear for each objective. If you cannot meet an objective, the page number will guide you back to the text that needs to be reread.

SUMMARIES

There is a summary at the end of each chapter and chapter section. The summary is designed to provide a quick review of the major points. Using the summary will help you organize your thoughts and will serve to alert you to information you may have missed or forgotten.

There is a summary system built into the entire book. Much of the important material is repeated in several places within the text. This has been done to give emphasis to certain facts and concepts, show you how to apply facts that were learned earlier in your readings, and to review information that may have been "lost" because it was read weeks ago. If something is presented more than once, IT'S IMPORTANT.

This system includes key fireground figures which are located in the "Notes Column." Their presence indicates major blocks of information devoted to the functions of the FGC, Company Officer, Sector Officer, Engine Company, Ladder Company, Rescue Company, etc.

COMMAND DEVELOPMENT AND REPORT CARDS

At the end of each chapter and chapter section, you will find material entitled "COMMAND DEVELOPMENT." These materials are offered to guide you in your course and during your experiences as a Fireground Commander. They are not to be taken as the "secret to success." Their whole purpose is to tie in what you have learned to basic fireground operations.

The "Report Card" section is added to help you keep track of your progress, chapter by chapter. If you cannot do something in Chapter 4, the report card will indicate this, allowing you to do the needed review before moving on. Learning will be more effective if you can apply previously learned materials to the next chapter.

USING THIS BOOK

To be an effective Fireground Commander, you need to pay special attention to the first three chapters. Chapter 1 defines the Fireground Commander. It tells you who he is and what he does. Chapter 2 explains how command is based on standard procedures. A Fireground Commander is expected to do certain things in a certain way. Chapter 3 covers all the major fireground functions of Command.

As you read each chapter, you should:

1. Read the list of objectives, making sure that you understand each one before reading the chapter.
2. Read the chapter, noting when and where each objective is covered.
3. Give special attention to each illustration, chart, and form presented in the text. They emphasize important facts and concepts. The cartoons are important. Each points out an important Command concept.
4. Use the summary to recall the major points of the chapter.
5. Go back over the list of objectives. See if you can do what is asked.
6. Reread the parts of the chapter that cover the objectives you cannot meet.
7. Study the information on command development to see if you

can relate the chapter to the role of the Fireground Commander.

8. Use the report card to follow your progress in exercises and simulations, and during fireground operations.

THE WORKBOOK

There is a workbook for this text. It is designed to help you meet the objectives for each chapter and chapter section. Use the questions and self-tests to be certain that you have mastered a chapter before going on with your reading. The workbook is self-instructional. It contains answers and references to the textbook.

1
THE FIREGROUND COMMANDER

MAJOR GOAL

TO DEFINE THE ROLES, RESPONSIBILITIES, AND FUNCTIONS OF THE FGC.

OBJECTIVES By the end of this chapter, you should be able to:

1. Define "Fireground Commander." (p. 2)
2. List the seven predictable areas in which fireground operations will break down if there is no FGC. (p. 2)
3. Explain how the designation of an FGC can help prevent this breakdown. (pp. 2-4)
4. Explain what is meant by "the FGC works at the strategic level rather than the task level." (p. 4)
5. List the four major responsibilities of the FGC. (p. 4)
6. Explain the role and responsibilities of the FGC in regard to the overall fire plan (p. 4)
7. List three things a FGC must be trained in so that he can carry out his responsibilities. (p. 5)
8. Explain the importance of open-ended decisions. (p. 5)
9. Describe how information is best gathered and organized by the FGC. (p. 6)
10. List and define four special management situations faced by most FGCs. (p. 8)
11. State the importance of review and evaluation. (p. 8)
12. Describe the traits of a good FGC. (p. 10)
13. List the eight major command functions of the FGC. (p. 11)
14. State why fireground etiquette is an essential part of all standard fireground operations. (p. 11)

1

COMMAND

THE NEED FOR COMMAND

An effective fireground operation centers around one incident commander. If there is no command, or if there are multiple commands, fireground operations quickly break down in seven predictable areas:

- Action
- Command and Control
- Coordination
- Planning
- Organization
- Communications
- Safety.

Action

There are times when firefighters do not take the correct action, doing things that do not follow the standard rules and principles of firefighting (e.g., aiming an attack line down through a roof vent hole or attack from the wrong direction). Such actions may endanger lives and result in an expanded loss of property.

Solution: Have a single commander structure action around tactical guidelines and see that all personnel follow the rules.

Command and Control

Fires with no command or having multiple commanders tend to produce chaotic, ineffectual action.

Solution: Strong, early, visible command by one individual who is responsible for controlling the entire operation will correctly mobilize the entire team. This requires a command system that designates one incident commander and provides him with the organizational support needed for correct command and control.

Coordination

When companies are not integrated under a central "game plan," they will quickly engage in independent actions. These actions seldom consider the collective capacities of the entire attack team. This *"free-lance firefighting"* will often work at cross purposes with the actions being taken by other units on the scene.

Solution: All tasks must be coordinated through a single incident commander. This person should establish the overall plan, assign companies to specific tasks, and assist companies in achieving their tasks by using effective direction of the operations. The goal of this commander is to get the maximum productivity from all available resources working together.

Planning

Effective firefighting requires a plan that is based on a prediction of where the fire is going and what it will do next. If there is no Commander, there will be no plan and no updating of the plan. If there are multiple Commanders, the plans generated will not be properly coordinated or efficiently updated.

Solution: Have a single Incident Commander who will combine an effective pre-fire planning system, reconnaissance, and information processing on the fireground to develop and update one plan. This person should formulate the strategic plan based on experience and an understanding of fire behavior, tempered by an appreciation of the fire environment (structural layout, construction, exposures, concealed spaces, fuel load, and protection features).

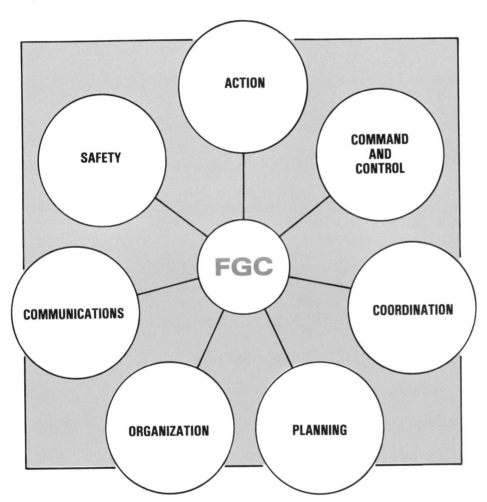

FIGURE 1.1: The designation of a FGC will help the breakdown of fireground action in seven areas.

Organization

With no overall game plan, it is doubtful that the participants will play their proper roles. This role confusion means uncoordinated action, resulting in the breakdown of the strategic, tactical, and task levels of firefighting.

Solution: Have a single Commander develop the plan and establish the roles, relationships, and functions for everyone on the fireground.

Communications

Communications difficulties are usually a reflection of other organizational problems that occur during fireground operations. Companies need to manage their own jobs and to exchange enough information with other companies to complete their tasks. When this information flow is slowed, over used, or stopped, confusion begins.

Solution: Have a single Commander who will use a standard operating procedure that supports command, coordination, and organization. Part of this procedure should be a communications plan that describes fireground information flow for all levels of operation.

Safety

Uncontrolled firefighting often leads to unnecessary, preventable firefighter injuries. All injuries have a detrimental effect on the entire operation, as well as their obvious effect on the victims.

Solution: Have a single Commander use a standard safety operating procedure to command, manage, and control the positions and functions of all companies at the scene. He must always reinforce safety.

ROLES AND RESPONSIBILITIES

The Fireground Commander is the individual with overall responsibility for incident command. The role of the Fireground Commander is one of a professional manager and commander. The term *"professional"* refers to training, dedication, and the desire to perform to the best of one's abilities and composure. It has no bearing on whether the FGC is a career officer or a volunteer.

The FGC's role as manager requires him to direct all fireground operations. As a Commander, he is expected to choose command over action, working from *strategic levels* rather than the task level. The FGC always moves toward the correct action, recognizing the correct thing to do, and knowing how to have others do it. While carrying out his role, the FGC is expected to operate in a clinically calm manner. He is expected to look and act as a professional at all times on the scene.

The four major responsibilities of the Fireground Commander are to:

1. Provide for firefighter safety and survival
2. Protect, remove, and provide care for endangered occupants
3. Stop the fire
4. Conserve property during and after fire control operations.

This list is also the same list of responsibilities for the fire department. Even though all personnel on the fire scene represent the fire department, the FGC has the responsibility for all activities on the fireground. Thus, his mission is the fire department's mission; his responsibilities are those of the department.

Specifically, the FGC's responsibility is to develop an overall plan related to a standard set of rules and principles—*the standard operating procedures* (SOPs). The development of this plan requires the FGC to anticipate and forecast outcomes based on his evaluation of conditions. During the operations, he is expected to review, evaluate, and revise this plan.

FIGURE 1.2: The FGC is the professional manager and commander at the fire scene.

So that his plan is developed properly and carried out correctly, the FGC is expected to manage risks, develop effective communications, develop effective organizations, and eliminate confusion. Throughout all phases of the operation, he is expected to maximize the use of resources to get the most out of personnel and apparatus.

The FGC has to evaluate operations during the fire, correct problems, and support effective actions. After the fire, he must evaluate all operations so that he can improve his personnel, his tactics, the system, and himself. He must strive to make sure that any future mistakes will be new mistakes.

In order to carry out his responsibilities, the FGC must be well trained in:

- Decision making
- Command and control
- Review and evaluation.

Decision Making

Decision making should be used on the FGC's realization that all tactical situations have the same basic elements and, therefore, he can apply a standard approach to them.

The FGC's decision-making process begins with the Commander knowing that he must avoid "dead-end" decisions. Whenever possible, his decisions must be *open-ended,* allowing for expansion, reversal, and building upon. Once his mind is set in this direction, the FGC must quickly establish the plan of attack and initiate action. The longer he waits, the less options he will have available.

Having to make quick decisions may worry the beginning Fireground Commander but they become easier to make once he learns that he has to:

DISTINGUISH BETWEEN ASSUMPTIONS AND FACTS— Operations must sometimes be based upon *assumed information*. Factual information is often incomplete. The FGC must realize that both the information and his decisions will have the opportunity to improve as the incident grows older.

MAINTAIN A FLEXIBLE APPROACH TO DECISION MAKING—The FGC should realize that he has the opportunity to update his plan and other decisions by utilizing *feedback* during the operations. Feedback allows for revisions to the general approach, specific tactical positions, and all major decisions.

DEVELOP A STANDARD RESPONSE TO REPORTED AND VIEWED CONDITIONS—Some basic facts and observations are needed to go along with the assumptions used to make initial decisions. By obtaining the required facts and applying standard responses, the FGC can avoid making premature decisions.

SHIFT TO A MANAGEMENT ROLE AFTER INITIATING ACTION—An FGC cannot make ALL ongoing fireground decisions. The efficiency of command decisions will improve once the FGC delegates tactical responsibility.

The FGC must quickly *prioritize problems* and *develop solutions*. This requires the effective gathering, recording, and organizing of information. Fireground intelligence can rapidly supply Command with random data and information. The FGC has to use a routine of information gathering and processing that is within his own mental limits. Without effective information gathering and utilization, overload quickly occurs and there can be no decision making.

Information is best gathered and organized by:

SEEKING OUT INFORMATION THAT IS CURRENT, ACCURATE, AND SPECIFIC—This requires direction on the part of the FGC.

USING DELEGATED INFORMATION RETRIEVAL—This keeps the FGC from having to depend too much on his personal view of the scene.

KNOWING WHERE TO FIND REFERENCE INFORMATION AND HOW TO USE IT EFFECTIVELY.

ASKING FOR AND GETTING THE RIGHT INFORMATION IN THE SEQUENCE NECESSARY TO COMPLETE THE ORDERED PRIORITIES.

UTILIZING A WIDE VARIETY OF FACTORS AND INFORMATION—This gives a "panoramic" view of the scene.

Command and Control

There is always some apprehension about taking command, especially for the new FGC. The development of confidence in himself and his personnel and the correct attitude soon make most FGCs anxious to accept responsibility and assume command, even during difficult operations.

Experienced FGCs regard the fire in enemy-oriented, pessimistic terms. They then apply command and control efforts to *achieve results,* not simply for the sake of taking charge. Confident FGCs refuse to be overwhelmed as they assume command. As soon as possible, they delegate certain responsibilities and insist that everyone make the decisions and do the job they have been assigned.

Successful FGCs apply flexible control levels, dependent upon conditions. They want to be able to order "what and where" without having to decide "how." Even though they do not wish to be slowed down by too many details, these FGCs realize that a few minutes spent establishing effective command at the beginning may save hours in the course of the operation.

Experienced FGCs know that they must select an appropriate command post and stay there. Their role involves strategic and tactical responsibilities, not task-oriented ones. They analyze a tactical situation in clinical terms, refusing to be distracted by visual conditions and refusing to rely only on what they see to make all decisions. They have delegated responsibilities, requiring others to keep them updated, and they are at the command post if a decision is needed.

Firefighting is a team effort. Well-trained FGCs delegate responsibility and seek feedback. They practice the art of "selective democracy." They can distinguish between the times to "call for a vote" and when not to, always realizing that their "vote" outweighs all the others.

FIGURE 1.3: The FGC must delegate responsibility.

Respected FGCs are those who attempt to develop realistic expectations for all officers and companies. They realize the differences in capacity, motivation, intelligence, and experience of their personnel and seek to place people in the best spot in order to get the most from them.

FGCs build a command support structure. They respect the command process, working with other operating commanders to strengthen the command function. Through support and cooperation, mixed in with a little understanding of the other person's problems, effective FGCs help eliminate the *"Let's put it out before the Chief gets here and screws it up"* syndrome.

There are four management situations that you must consider before taking command at the fireground. Unless you are prepared, your command and control may not be effective. These situations are:

STRESS MANAGEMENT—You must develop the ability to divide an overall problem into its parts and then delegate authority. This reduces the number of subordinates you will have to deal with, makes for easier control, and reduces the stress placed on you. It also reduces the stress on your officers by limiting their responsibility to the assigned aspects of the operation.

"LONE RANGER" MANAGEMENT—You must be ready to act as the single Commander, particularly in setting up the operation during the initial stages. True, you will delegate responsibilities, but you must take the ultimate responsibility for all fireground operations. This is not a problem once you learn to use other people on tasks that allow them to function at their highest level.

MIDPOINT MANAGEMENT—You must be prepared to inherit ongoing scenarios. This means that you must be prepared to evaluate the initial commitments and actions of others and make the necessary changes to fit your fire plan. Learn to do so with both confidence and grace.

SCARCE RESOURCE MANAGEMENT—You must be prepared to allocate and manage when needs outstrip resources. This is often the case at the beginning of an incident. If the problem continues through the incident, you will have to know how to obtain the needed support or how to use what you have in a way that protects personnel, victims, and property.

If you follow the advice of successful FGCs, you should be capable of establishing command and control. Problems will occur, but you and your officers will be able to solve them.

Command and control require a professional attitude—knowing what to do, when to do it, and how to motivate others to do their best. This is how you create your image as the FGC.

Review and Evaluation

Command and control will not be effective for an incident, or in the future, unless there is the ongoing process of *review and evaluation*.

Constant reassessment and possible revision of tactical operations are needed to maximize effectiveness. The FGC must be able to integrate evaluation and revision into the overall management approach.

The FGC needs to have a preoccupation with effective results, which means managing and evaluating with a focus on performance. He must review and evaluate with a high set of standards, expecting a high performance level from subordinates. He has to be intolerant of substandard output, reacting to correct what is wrong.

Review and evaluation are useful on the fireground only when the plan is kept open-ended. The FGC needs to plan ahead and operate with a back-up. He has to ask himself, "What do we do next if what we are doing now doesn't work?" If he sticks to a standard set of rules, principles, and priorities, a review of what he is doing might produce a better deployment of personnel and apparatus. The FGC must have his next step planned before he can order such changes.

The FGC has to apply a *pessimistic, critical approach* to the review of vital fireground elements. Basically, he should not believe that the fire is going to give up without a fight. He has to be willing to question reports from others and to seek confirmation. He must be prepared to disagree with a decision or countermand an order, doing so in a constructive manner. Part of this approach is founded on the principle that there are reporting limitations. Reports generally account for only the conditions as observed in the immediate area of the reporter. The FGC must consider the positions of the reporters inside and outside and the capability of the person giving the report to be fully aware of the situation.

The FGC cannot accept a bad situation. He has to be willing to admit that a mistake has been made or that conditions have changed and correct the commitment of personnel and apparatus. To do so, he needs to consistently utilize progress and condition reports for the purpose of revision and reinforcement. Unless the FGC is willing to make strategic and tactical revisions based on review and evaluation, his fire plan will be at a great disadvantage.

On the fireground, things go right and things go wrong. The FGC should evaluate the effectiveness of behavior, ability, and performance in relation to rank. The higher the rank, the more critical should be the review. He should never expect anything from subordinates that they have not been trained, prepared, and equipped to do.

The FGC must assume a *positive, supportive leadership role* when things go wrong. The need for quick, effective action requires that teaching and discipline take place before the actual event. The FGC and his officers must work to correct dysfunctions when they occur to produce the desired results.

The fireground operation is not over until the FGC has performed a postfire critique. This should be a learning tool that will give credit to those deserving it and correct substandard performance. The critique should reinforce superior performance and leave personnel with a positive feeling. Constructive criticism should always be directed with a positive feeling. Constructive criticism should always be directed only to the individual involved. There is no reason for spectators inside and outside of the department to take part in negative performance reviews. To do otherwise will eventually erode your command as bad feelings develop between you and your personnel. If direct behavior modification actions are necessary, they should be managed in an appropriate manner and setting.

Most important of all, no evaluation and review approach can be effective unless the leader also accepts criticism. The FGC must welcome review, revision, and constructive criticism as well as be willing to give them to others.

FGC TRAITS

Command on the fireground is a function of the ability and philosophy of the person in command. The personality of the FGC is critical to command and control. Desirable traits for a FGC include:

RESPECT FOR THE TASK—Understanding that victims are rescued and the fire is controlled by firefighters who are generally doing difficult tasks

ABILITY TO STAY COOL—Command composure allows the FGC to successfully manage operations and maintain the confidence and respect of his firefighters

KNOWLEDGE OF COMMAND—Developed through training and experience.

AN INCLINATION TO COMMAND, NOT ACT—Able to make decisions and have others do the tasks

ABILITY TO PROVIDE A POSITIVE EXAMPLE

BEING PSYCHOLOGICALLY STABLE—Able to remain clinical as he regards the fire scene and its possible outcomes

BEING PHYSICALLY FIT—Able to endure the time and stress of the incident

FAIRNESS—Realizing that command is not a popularity contest but respecting the abilities and feelings of others

BEING STRAIGHTFORWARD WHEN COMMUNICATING

WILLINGNESS TO TAKE REASONABLE RISKS—Without compromising safety

CONCERN FOR ALL PERSONNEL

KNOWING LIMITATIONS—Of himself, others, apparatus, and his strategic and tactical approaches

RESPECT FOR COMMAND

BEING AN ORGANIZATION PERSON

BEING DISCIPLINED AND CONSISTENT

COMMAND FUNCTIONS

The majority of the roles and responsibilities of the FGC are carried out through a series of specific command functions which form a FGC job description. Chapter 3 covers each of these functions, one per section. Read over the list and see how your concept of the FGC's roles and responsibilities can be carried out through the application of these functions.

The command functions of the Fireground Commander include:

1. Assumption, confirmation, and positioning of command
2. Situation evaluation
3. Initiation, maintenance, and control of the communications process
4. Identification of the overall strategy, development of an attack plan, and the assignment of units
5. Development of an effective fireground organization
6. Reviewing, evaluating, and revising the attack plan
7. Providing continuing command, transferring command (as required), and terminating command.

Fireground Etiquette

Effective fireground operations are highly influenced by fireground etiquette. This refers to the manner in which operations are accomplished on the incident scene. Too often, there is a breakdown in this etiquette because of fire scene excitement, stress, and a clinical emphasis on the most intricate details of the SOP. Avoiding these etiquette violations leads to healthier, happier, and more effective operations. The FGC should:

BE A LEADER—The leader supports the group and the group supports the leader. Do not debate, argue, or vote to see who is "correct."

RESPECT PERSONNEL—Be sensitive to their needs. Forget personalities, politics, and personal quirks. Work with everyone to beat the fire. Do not ridicule or berate someone who makes a mistake. Handle it properly during the critique.

NOT TAKE ADVANTAGE OF RANK, AUTHORITY, OR SENIORITY—Everyone does his share of work, with the able supporting the less able. If someone is in trouble, help him. If he has made a mistake, help him correct it. If you think he is about to make a mistake, help prevent it.

ELIMINATE MULTIPLE STANDARDS—Make certain that everyone plays the same rules. Do not play favorites and do not try to get even. Never use assignments as punishment.

NOT WASTE TIME WITH "FIREGROUND HOBBIES" — Every fire officer has a pet theory or favorite tactic. Use them when they are appropriate, but don't try to make every situation fit the same tactic.

EXTEND A REASONABLE DEFERENCE TO RANK AND SENIORITY—Respect the people who are running the show. There is no reason, or time, to salute on the fireground.

USE PROPER LANGUAGE—Be professional in the way you give orders, seek information, and receive reports. All conversations and other communications should be done with language and courtesy that is appropriate for the public record.

SUMMARY

There must be one commander for a fireground operation or there will be a quick breakdown of action, command and control, coordination, planning, organization, communications, and safety.

The FGC is the person who assumes overall command of personnel and apparatus at the emergency incident scene. He is a manager employing special management systems designed to consider the danger, compressed time, incomplete and inaccurate information, difficult communications, and confusion on the fireground.

The role of the FGC is to be a professional manager and commander, directing all fireground operations. He must choose command over action and carry out his functions at the strategic and tactical level rather than the task level. It is essential that the FGC move toward the correct action, knowing the correct thing to do and how to have others do it.

The FGC is responsible for providing for the safety and survival of all personnel; protecting, removing, and providing care for all endangered occupants; stopping the fire where it is found; and conserving property during and after fire control operations.

The FGC must develop an overall plan related to standard operating procedures. This plan should anticipate and forecast outcomes. It must be reviewed, evaluated, and revised during operations.

To carry out his functions, the FGC must be well-trained in decision making, command and control, and review and evaluation.

Decision making at the fireground is based on the principle that *all tactical situations have the same basic elements requiring the application of standard approaches.* Fireground decisions must be open-ended, allowing for expansion, reversal, and building upon. The plan must be quickly established and action initiated as soon as possible. This requires the FGC to quickly prioritize problems and develop solutions. Information must be effectively gathered, recorded, and organized if the FGC is to perform this function.

Information must be current, accurate, and specific. The FGC should delegate information retrieval, knowing where to find reference information, asking for the right information in the correct sequence, and utilizing a wide variety of factors and information for decision making.

Command and control efforts should be applied to achieve results. The FGC should apply flexible control levels with responsibilities properly delegated. He should stay at a fixed command post and manage, carefully using others to provide him with information, observations, and opinions. The FGC must be familiar with stress management (delegating responsibility), "Lone Ranger" management (taking action and assuming overall responsibility as the single commander), midpoint management, and scarce resource management.

Review and evaluation must be an ongoing process in order to maximize effectiveness. This must be done with a high set of standards, expecting high performance levels. The FGC must react to what is incorrect and provide a solution. The next step must always be planned in case something does not work. The FGC must be willing to admit that a mistake has been made or conditions have changed so he can modify the plan and correct the action being taken. He should assume a positive leadership role when things go wrong. Constructive criticism should be provided and a post-fire critique must be done.

All review and evaluations should leave personnel with a positive feeling. Privacy is required for individual criticism.

The FGC's personality is a big factor in the command system. The desirable traits for a FGC include the required knowledge of command and the inclination to command, control of temper, the ability to provide a positive example, psychological stability, physical fitness, fairness, being straightforward when communicating, a willingness to take reasonable risks without compromising safety, concern for all personnel, knowledge of limitations (self, personnel, apparatus, the plan), respect for command, being an organization person, and being disciplined and consistent.

Fireground etiquette reflects the manner in which fireground operations are accomplished. The smooth flow of operations requires the FGC to have a realistic view of rank, respect for personnel and their tasks, to be a leader, and to be fair.

The FGC's roles and responsibilities are carried out through a specific set of command functions (see Chapter 3 for complete details).

COMMAND DEVELOPMENT

To be a successful Fireground Commander, you must develop your skills in decision making, command and control, and review and evaluation. To be effective, you must have a working knowledge of fire and fireground operations. You must study what to do and when to do it. You must understand how something is done so that you will know if it is practical for a given situation and how much manpower and time will have to be committed. Your knowledge and skills must be used to motivate your subordinates to do the tasks necessary to carry out your overall plans. This will develop as you evolve as a professional FGC and as you gain more practical experience.

The following report card is provided so that you can evaluate your FGC knowledge and skills in classroom exercises, simulations and on the fire scene.

Fireground Commander Report Card

Subject: The Fireground Commander

Did the Fireground Commander:

- ☐ Effectively manage and control the seven areas of predictable breakdown?
- ☐ Use quick, efficient decision making?
- ☐ Properly delegate responsibilities?
- ☐ Keep his plan open-ended?
- ☐ Work at the strategic level?
- ☐ Update information and change his plan as needed?
- ☐ Review and evaluate all fireground operations?
- ☐ Properly give credit for outstanding operations and privately offer constructive criticism for substandard performance?

2

STANDARD OPERATING PROCEDURES

MAJOR GOAL

TO DEVELOP STANDARD OPERATING PROCEDURES FOR FIREGROUND OPERATIONS.

OBJECTIVES By the end of this chapter, you should be able to:

1. Define "standard operating procedures." (p. 16)
2. State the purpose of a fire control system model. (p. 19)
3. State the major elements found in all SOPs. (p. 16)
4. List at least five types of SOPs required for managing the fireground. (p. 16)
5. State the four categories of fireground operations requiring SOPs. (p. 19)
6. List the participants of the typical fireground action and state their roles. (p. 20)

STANDARD OPERATING PROCEDURES

STANDARD FIREGROUND PROCEDURES

Standard fireground procedures are a set of organizational directives that establish a *standard course of action* on the fireground to increase the effectiveness of the firefighting team.

It is difficult for any fire department to operate consistently and effectively without such standard operating procedures (SOPs), particularly during large, complex, or unusual field situations. SOPs allow the organization to develop the *"game plan"* before the fire—one of the most important elements of prefire planning.

Successful fireground command and firefighting activities require the integrated efforts of the entire team which is, in turn, organized and mobilized under a strong central plan. While the actions of individuals on the fireground are important, firefighting operations are done collectively.

Obviously, each department must develop procedures that apply to particular problems. Local conditions, capabilities, and limitations will define specific responses. However, our problems are more alike than different. A consistent management goal is to simplify, remove the mystery, and standardize operations to allow FGCs to begin to relate to one another. (Watch out for the guys who say, "We fight fire different out here,"—they generally do.)

The following material establishes a general framework and starting point for the development of SOPs tailored to fit local conditions.

Standard operating procedures should be developed to outline and describe an organizational approach to the major categories of fireground activity. They generally include such areas as:

- Basic command functions—including a standard method of assuming and continuing command
- A method to divide command responsibility through the delegation of areas and functions to sector officers
- All aspects of communications and dispatching
- Fireground safety
- Guidelines that establish and describe tactical priorities and related support functions
- A regular method of initial resource deployment
- An outline of responsibilities and functions of various companies and units.

SOPs are characterized by being:

- Written
- Official
- Applied to all situations
- Enforced
- Integrated into the management model.

Written

Unwritten directives are difficult to learn, remember, and apply. They tend to waste a lot of time and motion for everyone involved. The answer is to have officers and firefighters decide how all operations will be conducted and then commit those decisions to writing.

This documentation process and the agreement that goes with it removes the "blue sky" from fire functions.

The act of writing (difficult for most of us until we get started) requires the team to work their way through all natural confusion that accompanies the production of any standard. It is most difficult to write the script while the play is going on. Written procedures keep everyone honest by providing a set of players and an overall organizational plan developed before the fire. Procedures aren't procedures until they are written.

Official

SOPs become a collection of values and experiences that evolve into a fireground playbook and which represent the official policy of the organization. This manual also provides a convenient package for new members to learn from and experienced members to review the official fireground approach.

This system eliminates the game of trying to guess what will happen next on the fireground. It defines roles and responsibilities and creates an official structure where leaders lead and followers follow, according to the script.

A very real management problem is trying to get everyone to do standard operations the same way. Each crew, shift, district, and battalion tend to develop very individual and unique ways of doing the same thing. This fragmentation can create a number of small departments operating within the organization. It is virtually impossible to achieve department-wide improvement or consistency within this "subcontractor" world. The typical comment to the newly placed rookie is too often, "Forget how they do it at the Training Academy, this is how we do it 'out here' at Engine 81." Written SOPs give the organization a fighting chance to eliminate this fragmentation.

Applied To All Situations

The organization must make a commitment to use SOPs on all field activity all the time. Applying the guidelines to everyday, routine business develops a set of regular habits for the individual and the group.

If a variety of decisions have been made before the fire that structure how the organization will react, the FGC can concentrate on critical rather than routine decisions. He simply cannot do that if he has to decide where everyone will park their rig and how they should talk on the radio.

Enforced

Effective SOPs require a full range of activities, from planning to follow-up. How the command staff supports the system will determine how effective and durable it is.

Implementing a more organized system involves a major change from a more unstructured organization. Effective change of this type requires vigorous leadership to ensure that operations are really performed as they were written.

This enforcement should be educational, especially in the beginning. During these implementation periods, firefighters are busy trying to learn new methods and unlearn old habits, not an easy chore. The very best method of enforcement is acknowledgment and positive reinforcement of good performance. Recognition of positive experiences creates an atmosphere which motivates everyone to use the next opportunity to apply the procedure again.

A well-managed system that stresses open participation throughout the development and implementation phases eliminates the need for most negative discipline.

SOPs only effect what's happening on the fireground when they are carried out. A small, compact, practical set of enforced procedures will accomplish much more than a basket of complicated regulations that no one can remember. Leaders define themselves by what they enforce. Firefighters know what the boss will and won't stand for on the fireground. If the procedure is in the book but isn't being followed, enforce it, change it, or eliminate it.

Standard operating procedures become the basis for much of the use of the regular management process. The standard steps of system development/training/application/review/revision are used in the development, application, and ongoing management of SOPs.

THE FIRE CONTROL SYSTEM MODEL

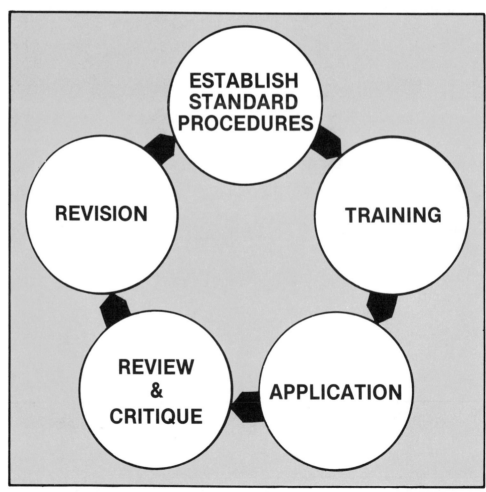

FIGURE 2.1: The fire control system model contains the activities required to manage SOPs.

Standard operating procedures serve as the foundation for a very simple *fire control system model*. This model defines the necessary management activities and the relationships between them. The ongoing operation of the system should continually improve fireground performance.

System Development

The first step in the model forces the organization to decide on its overall incident management philosophy. It requires the actual definition of each step in the process and, although all levels of the organization are involved, it must be supported and directed from the top. This plan must be defined in a set of written directives which are flexible and expandable to deal with routine situations and predictable escalations. This development process should begin by carefully analyzing current operations and deciding what is working well and what isn't. The object is to build on strengths and correct weaknesses.

Training

SOPs provide a practical training package for everyone. They allow members to become familiar with the system before it is actually used. Basically, *we should expect them to play the way they practice.* Practicing these procedures develops standard roles and functions within the team. The plan aligns the strategic, tactical, and task levels of the organization which are, in turn, directed toward everyone doing his job—Commanders command and firefighters fight fire. The entire system is now focused on effective action, mostly at the task level.

Field Operations

Actual fires are "show time" for the entire system. Field operations structure and solidify the "paper plan." The procedures should create a framework which will pinpoint the fireground objectives as *performance targets* (rescue/fire control/property conservation) and shift operations from minor to major and, eventually, back to minor. The FGC now has a set of directives that can be applied selectively to each particular situation.

Review and Critique

We know that fire operations test the system under actual conditions; the review element evaluates that field test. The standard procedures package becomes the basis for the postfire critique. The review should reinforce good performances and help resolve both individual and collective problems that may have occurred. For this to really happen, the organization must be willing to consider positive and negative outcomes and also to integrate them back into the SOPs. The critique now becomes an ongoing management tool that everyone expects to be a part of every incident. After the fire is the time to discuss and focus on lessons learned.

Revision

Changes in field conditions and departmental capabilities (e.g., closing a station) may require adjustments to the SOP. Procedures revision based on experience and forecasting has to be a regular element of management if it is to be successful. This "maintenance activity" is time consuming, but necessary, in order to stay current. The organization should develop a continuing capability to "fix itself" through the ongoing application of the Fire Control System Model.

FIREGROUND ROLES

Consider the *"cast of characters"* involved in firefighting. SOPs should predict the typical participation of each group and create a functional role for that participation. This anticipation of roles and functions becomes an advantage for everyone operating on the fireground. (The script must be written around the actual characters that came together in the play.) This basic fireground cast includes the following characters.

The FGC

The FGC is the person responsible for *managing the incident on the strategic level.* He establishes the overall operational plan, develops an effective organizational structure, allocates resources, makes assignments to carry out the attack plan, manages information to develop and revise decisions, and continually attempts to achieve the basic command objectives.

The FGC should establish and operate from a stationary command post as soon as possible. This command post should offer a relatively quiet vantage point, with lighting, radio equipment, and sufficient space to manipulate written reference material. The command system works because the FGC establishes the plan and everyone else follows it.

Sector Officers

Sector Officers are assigned by the FGC to manage specific geographic areas of the incident scene or specific fireground functions. They manage at a *tactical level* to achieve direct objectives within the overall strategic plan.

The FGC organizes the command structure by establishing Sectors and assigning Sector Officers to meet the management and operational needs of each specific incident. Sectors are created selectively—only the sectors that are required are assigned. (There is no advantage in assigning a Roof Sector to a fully involved structure.) When a sector assignment is made, the Sector Officer physically goes to the area, determines the resources needed, directly manages the companies to achieve the tactical objectives, and communicates with the FGC on the progress of the group. The FGC now has a method to *decentralize command* yet maintain a manageable span of control.

Command Aides

Aides are personnel assigned to assist the FGC. During large operations, Sector Officers also may have aides to assist them. They do this by managing information and communications. Aides can be any knowledgeable fireground operator—from a rookie to the Chief of an adjoining jurisdiction. They can keep track of assignments, locations, and the progress of companies, assist with tactical worksheets, or use reference materials and prefire plans. Another important function they may have is to provide reconnaissance and operational details for the commander (his eyes and ears). Some jurisdictions assign full-time aides to command officers to perform routine administrative functions and to act as drivers in addition to their fireground role.

Fire Companies

Fire companies are groups of workers who respond to the incident on apparatus. They carry out assigned tasks on the fireground. General classifications include engine, ladder, rescue squad, and special service companies. These companies work under the direct command of the FGC or designated Sector Officers. Each company should have a Company Officer who supervises the work and is responsible for maintaining communications with the command structure.

Dispatchers

Dispatchers provide the central communications function for the fire department. They receive calls from citizens requesting emergency services and then dispatch the appropriate units to the correct location. In addition, they continue to support the operation by providing a communications link and assist the command system by dispatching and coordinating the response of any additional resources requested. Dispatch is a critical command system element.

Support Personnel

Support personnel represent areas of responsibility from within the fire and rescue services and from outside agencies that routinely respond to the firm alarms. They provide services which are often critical to firefighting operations. For example, they can provide utility control, water supply, mechanical repair, special equipment, technical consultation, and welfare services. It is important that these services be integrated into the overall command system.

Fire Investigators

Investigators are responsible for determining the cause and the origin of the fire. They respond to the fireground and integrate their efforts with the overall operation. Their combination of firefighting experience and skills, combined with the investigative authority of police officers, will usually provide the final link in the fire protection system by determining how the fire actually happened.

Victims

Fire victims are the people who may be injured or killed as the result of a fire or those who are displaced because of property damage.

When a fire occurs, they must depend on the skills, abilities, and organization of the responding services. The command system should always be oriented toward the victims and geared to minimize the physical and psychological impact of a fire.

News Media

Fires are certainly significant news events, and the news media attend fires to report those events. Media employees are typically intelligent, aggressive people who understand their obligations and constitutional right to inform the public. Their efforts comprise the normal methods used within the community to inform the public. Positive, well-written accounts of fire operations can have a major impact on the public image of the department as a whole.

Nevertheless, facing reporters on the fire scene can be a distraction, to say the least. Procedures must be built into the system to accommodate their reporting needs. A "Public Information Sector" should be established during department operations to provide essential information to reporters and to insure their safety. One person, dealing with the media as a group, can ensure that all reporters get the same facts, without interrupting the FGC.

Spectators

Spectators are curious members of the general public who visit the incident scene to watch the fire and firefighters. They only become a problem for the command system when they interfere with operations or become exposed to fire hazards. Spectators are often difficult to manage and control. Since the FGC is responsible for the welfare of everyone on the fireground, he must include their safety and control into his overall plan.

Police

Police represent the community agency with the authority and ability to directly control the location and activity of the general public at an emergency scene. This capability makes them a unique support agency for the fire command system through their ability to control and manage spectators, traffic, and other actions of people. The command system should integrate law enforcement functions into its operation as a matter of routine.

The Fire

Any realistic fireground "cast of characters" must include the villain—the fire. The fire is the major problem and is the central challenge for the rest of the cast. No one can ever afford to regard the fire as an inactive, impotent, or routine enemy. To be effective (and to survive) the actors must develop a practical respect for the fire and an understanding of its "tricks."

Staging LEVEL I	**PHOENIX FIRE DEPARTMENT** **STANDARD OPERATING PROCEDURES** M.P. 203.02 08/84-R Page 1 of 2

LEVEL I — STAGING

Level I Staging will automatically apply to all multiple unit responses unless otherwise ordered by Command. Level I Staging involves the following:

> **The first arriving engine company** will respond directly to the scene and will operate to best advantage.

> **The first arriving ladder company** will respond directly to the scene and will place themselves to best advantage, generally at the front of the building and report their action by radio.

> **The first arriving rescue unit** will go directly to the scene and place their apparatus in a location that will provide maximum access for medical/rescue support and not impede the movement of other units and indicate their action by radio.

All other units will stage in their direction of travel, uncommitted, approximately one block from the scene until assigned by Command. A position providing a maximum of possible tactical options with regard to access, direction of travel, water supply, etc., should be selected.

All engine companies will pull map books and reference immediate fire area for water supply.

Staged companies or units will, in normal response situations, report company designation, standing by and their direction ("Engine One, South"); it may be necessary to be more specific when reporting standby positions in extraordinary response situations. An acknowledgment is not necessary from either Alarm Headquarters or Command. Staged companies will stay off the air until orders are received from Command. If it becomes apparent Command has forgotten the company is in a staged position, the company officer shall contact Command and readvise him of their standby status.

These Staging procedures attempt to reduce routine traffic, but in no way should reduce effective communications or the initiative of officers to communicate. If staged companies observe *critical* tactical needs, they will advise Command of such critical conditions and their actions.

When arriving at staging, companies will indicate their status as "Staged" by MDT. If assigned to a task, they will indicate "On-Scene" status.

FIGURE 2.2: An example of Level I Staging SOPs.

Staging LEVEL II	**PHOENIX FIRE DEPARTMENT** **STANDARD OPERATING PROCEDURES** M.P. 203.03 08/84-R Page 1 of 3

LEVEL II — STAGING

Level II Staging is used when an on-scene reserve of companies is required. These companies are placed in a Staging Area at a location designated by Command. When Command announces "Level II Staging," all 2nd Alarm and greater companies will report to and remain in the Staging Area until assigned. First alarm companies will continue with Level I Staging unless instructed otherwise. When going to Level II, Command will give an approximate location for the Staging Area. Companies which are already staged (Level I) will stay in Level I Staging unless advised otherwise by Command. All other responding units will proceed to the Level II Staging Area.

The Staging Area should be away from the Command Post and from the emergency scene in order to provide adequate space for assembly and for safe and effective apparatus movement.

When calling for additional resources, Command should consider Level II Staging at the time of the call. This is more functional than calling for Level II Staging while units are enroute. The additional units will be dispatched to the Staging Area.

Command may designate a Staging Area and Staging Officer who will be responsible for the activities outlined in this directive. In the absence of such an assignment, the first Fire Department officer to arrive at the Staging Area will automatically become the Staging Officer and will notify Command on arrival. The arrival notification will be made to Command on the assigned tactical channel. (Command may direct Staging to communicate on Channel 5.)

Due to the limited number of ladder companies, a ladder captain will transfer responsibility for Staging to the first arriving engine company captain. Staging Officers will assign their company members to best advantage.

In some cases, Command may ask the Staging Officer to scout the best location for the Staging Area and report back to Command.

The radio designation for the Staging Officer will be: "STAGING." All communications involving Staging will be between Staging and Command. All responding companies will stay off the air, respond directly to the designated Staging Area, and report in person to the Staging Officer. They will standby their unit with crew intact and warning lights turned off. Staged units will indicate their status via MDT as "Staged."

FIGURE 2.3: An example of Level II Staging SOPs.

SUMMARY

A standard course of action on the fireground can be established by using standard fireground procedures. SOPs allow the organization to develop a "game plan" before the fire.

Firefighting operations are performed collectively. A strong central plan will integrate the efforts of the entire team.

The *fire control system model* defines the necessary management activities and details the relationships between them. The four categories of operations that require SOPs are training, field operations, review and critique, and revision.

The first step of the model decides the overall incident management philosophy. Each step or phase of the process is defined. These standard procedures provide everyone with a practical training package used to develop standard roles and functions and to focus on effective team action at the task level.

On the fireground, the FGC can use these procedures for each situation. The framework of objectives and targets of operations have been built before the fire.

Adjustments to standard procedures start with the review and critique used to evaluate past performance and identify problems. Revision is done as a maintenance activity based on the fireground experience, the review and critique, and forecasting.

Every fireground operation has a cast of characters. When developing SOPs, each of their roles must be defined. The typical cast should include: *the FGC*—responsible for managing the incident on the strategic level; *Sector Officers*—commanders assigned by the FGC to specific geographic areas of the scene or to specific fireground operations; *Command Aides*—personnel assigned to assist the FGC; *fire companies*—groups of workers who respond on apparatus; *dispatchers*—provide central communications; *support personnel*—provide services critical to the operation; *fire investigators*—responsible for determining the cause and origin of the fire; *victims; news media; spectators;* and *the police.* Remember that the villain in the cast is the fire.

COMMAND DEVELOPMENT

As a FGC, you must work hard with your system to develop standard fireground procedures. You must realize that it will be difficult to operate effectively without these standards. The successful FGC is able to organize and mobilize his personnel based on prefire planning. Standard procedures are the most important part of this planning.

The system is dependent on the FGC to field test procedures, refine definitions of fireground personnel roles, and identify needed standard procedures. As an FGC, you will be expected to prepare and enforce such changes.

The following report card is provided so that you can evaluate your FGC knowledge and skills in classroom exercises and simulations and on the fire scene.

Fireground Commander Report Card

Subject: Standard Operating Procedures

Did the Fireground Commander:

- ☐ Have a working knowledge of his department's SOPs?
- ☐ Use SOPs to help pinpoint fireground objectives?
- ☐ Review and critique the SOPs used?
- ☐ Know the roles of his personnel?
- ☐ Utilize SOPs throughout the incident?

3 FUNCTIONS OF COMMAND

Section 1:
ASSUMPTION, CONFIRMATION, AND POSITIONING OF COMMAND

MAJOR GOAL

TO QUICKLY ESTABLISH AND CONFIRM A SINGLE FGC AND PLACE THAT INDIVIDUAL IN THE MOST EFFECTIVE COMMAND POSITION.

OBJECTIVES By the end of this section, you should be able to:

1. State why central command is needed on the fireground. (p. 28)
2. List two typical fireground outcomes which may occur when there is no effective command system. (p. 28)
3. List three objectives of the command procedure. (p. 28)
4. State the major component of the command procedure. (p. 29)
5. Describe the process of assuming command. (p. 30)
6. List and define the three command modes for a Company Officer initially assuming stationary command (p. 30)
7. Describe the relationship of company arrival time to the Company Officer's decision to assume command. (p. 32)
8. List and describe the three ways in which a Company Officer may commit his company when assuming command. (p. 32)
9. Describe how the FGC announces his assumption of command. (p. 32)
10. List the three categories for personnel who arrive at a scene having a designated FGC. (p. 33)
11. Describe the standard position for the FGC at the incident scene. (p. 33)
12. List five major advantages of a stationary command post. (p. 33)

CENTRAL COMMAND

THE NEED FOR CENTRAL COMMAND

Successful fire operations require the application of an effective overall management system and the skills of a strong Fireground Commander. The lack of understanding of this central command role adversely affects more fires than any other single management problem.

Simple, hard-hitting fireground operations are conducted at the task level and require a strategic plan, application of SOPs, and continual management. Without *strong, central command,* the typical fire scene quickly deteriorates into an unsafe, out-of-control situation. It is highly unlikely that companies operating in this atmosphere will coordinate their activities and achieve the planned objectives.

Operating on the fireground without central command usually produces:

NO COMMAND AT ALL—Everyone operates in the "free enterprise system," where companies commit themselves independently with no central coordination.

MULTIPLE, COMPETITIVE COMMAND—The fireground quickly becomes occupied by several highly mobile command officers, each with a different plan and each wanting a piece of the action. They generally circle the fire giving conflicting orders to everyone they encounter. Portable radios give each orbiting commander the ability to communicate his confusion to everyone, without losing a step.

In both cases, the companies are usually made up of capable, dedicated people who are highly motivated to do a good job. Firefighting operations are important to them, and they deserve a chance to be successful. But, they need a good management system—one that attempts to apply just the right amount of command; too little creates uncontrollable chaos, while too much produces constricted action which underutilizes the skills of the players and the coaches.

ADVANTAGES OF CENTRAL COMMAND

The command procedure mobilizes and integrates everyone's efforts to:

1. Fix command responsibility on a particular individual through a standard identification system. The specific identity of the FGC depends upon the arrival sequence of firefighters, companies, and officers.
2. Ensure that strong, direct, and visible command is established as early as possible.
3. Establish a management framework that clearly outlines the objectives and functions assigned to the FGC.

This procedure defines the fundamental job of the FGC—establishing him as the *overall site manager.* His ongoing role is, then, to manage and coordinate all SOPs that collectively form the organizational structure.

The command procedure establishes one person as "Command." It identifies the person in command, the functions of command, and the location of command (command post). It does not, however, set up the FGC as a single figure on a white horse. Instead, it places Command in a position to use aides, staff officers, and other key personnel as command assistants when required. Well-managed command posts are generally not lonely places.

FIGURE 3.1.1: Command procedures have to be realistic.

Command procedures also give the FGC the responsibility for overall fireground results along with the authority to achieve those outcomes. This game plan balances the classic conflict between responsibility and authority and puts everyone to work on common objectives.

A major component of command procedures is, then, the agreement that the Fireground Commander will be the overall incident site manager and that everyone else operating on the fireground will follow and support his plan. This basic management agreement is particularly crucial in view of the urgent need for effective action on the fireground among all the participants.

Command procedures also establish a framework for reviewing fire outcomes (critiques). The standard review of procedures, training, application, and results through the critique produces a simple, straightforward management method that should continually improve the system's performance.

The *command functions list* provides a practical job description for the FGC and helps define what "being in command" really means. It also serves as an outline for everyone on the fireground to understand exactly what the FGC will be doing. A clear statement and agreement of the command functions serves as the common basis of the operational system for everyone.

The "job" of the FGC is a difficult one, with the normal challenges of managing the fireground producing many command performance

problems. While it is amazingly easy to criticize poor commanders, it is much more productive to take a positive approach to define what is expected, provide training in the basic functions, and use experience to improve ongoing performance through the educational review process. Current FGCs should be treated with kindness. These commanders are not the product of any management model—they were forced to learn the business the hard way. The challenge is to process the next generation through a better development system.

ASSUMPTION OF COMMAND

FIRST-ARRIVING FGC

Normally, the first-arriving unit or officer should be responsible for initially assuming command. This Fireground Commander retains command responsibilities until relieved of command by a higher ranking officer or until the incident is terminated. This assumption of command by the first unit should be mandatory; however, a Company Officer may choose to quickly pass command to another officer or unit under certain circumstances.

As the identity of the FGC changes through command transfers, the responsibility for command functions also changes. (The FGC is responsible for all command functions, all the time). The term *"Command"* now refers jointly to the person, the functions, and the location of command and provides a standard identification tag for the FGC. With this system, it should be all but impossible for more than one Fireground Commander to exist at any one incident scene.

COMMAND MODES

When the first-in officer arrives with a company, he must quickly decide which of the following commitments he should make:

1. **NOTHING SHOWING MODE**—generally requires investigation by the first-arriving engine or ladder (rescue unit) while other companies remain in Level I staging (all other units stage in their direction of travel, approximately one block from the scene or as conditions dictate). Usually, the officer will go with the investigating company while using his portable radio to continue the command function. In effect, he creates a "mobile command."

2. **FAST ATTACK MODE**—requires immediate action to stabilize, (e.g., a working, interior fire in a residence, apartment, or small commercial occupancy). For a critical offensive fast attack, the officer may choose to lead the attack while utilizing a portable radio to continue command, or he may transfer command to the second-arriving officer before entering the structure.

   This fast attack mode should be concluded rapidly with one of the following outcomes:
   a. Situation stabilized by offensive attack
   b. Command transferred to command officer (or another company officer)
   c. Situation not stabilized; company officer moves to exterior (stationary) command position.

3. **COMMAND MODE**—because of the size of the fire, complexity of the occupancy, or the possibility of extension, some scenarios will demand strong, direct command from the outset. In these cases, the Company Officer will initially assume a stationary exterior command position and remain there until relieved of command.

Even though the Company Officer who assumes command has a choice of command modes and degrees of personal involvement in the attack, he still remains responsible for the command functions.

While the FGC has the option of mobile command, he must realize that moving around and utilizing a portable radio places him at a disadvantage when performing his command responsibilities. This effective combination of initial action and command is a real test of initiative, attention span, and judgment under pressure.

The experienced Company Officer gives up the advantage of a stationary command post only when he realizes that his direct, personal participation will make a critical difference in the attack outcome or the safety of his crew. At this point, every officer on the fireground should recognize that the operation is in the *fast attack mode* and be prepared to support the initial attack effort.

FIGURE 3.1.2: A fast attack may require the direct participation by the FGC.

This "yes-no" mobile command decision is further influenced by the arrival time and sequence of responding units. When the next company is 10 minutes away, direct action is indicated. On the other hand, when the Company Officer arrives seconds ahead of the next five units, his efforts in coordinating the activities of all companies will probably be much more affective.

In cases when the first-arriving officer is a Command Officer, he should automatically assume a normal stationary, exterior command position and immediately begin the FGC functions. Command officers serving as the FGC do not have the option of mobile command.

PASSING COMMAND

While assumption of command by the first-arriving unit or officer should be mandatory, there are situations where it may be advisable to quickly pass command to another officer.

One example is when a command officer is arriving seconds behind the first company. Another would be where the initial-arriving company officer is faced with an urgent rescue, a critical tactical situation, or a particularly dangerous risk to his personnel and feels that his personal involvement is absolutely required. This is a true dilemma for the officer of an understaffed company.

The initial FGC is still responsible for all command functions until the other officer has arrived and assumed command.

CREW ASSIGNMENTS

When a Company Officer assumes a stationary command position, he has several options in regard to the assignment of his crew:

1. *"Move up" an acting officer within the company.* This is determined by the individual and collective capacity of the crew.
2. *Assign company personnel to staff (Aide) functions.*
3. *Assign company personnel to another company officer working under his command.* This must be acknowledged by both the original and the new officer.

CONFIRMATION OF COMMAND

The first-arriving unit or officer who will assume the role of Fireground Commander should advise Alarm of this fact by broadcasting the unit designation, arrival, assumption of command, and the name and location of that command post. For example, "Engine 1 on the scene, side 1 as Ajax Command."

Command may use the location of the incident (e.g., "Main Street Command") or the name of the occupancy (e.g., "Ajax Command") to designate that particular command operation. This naming system helps keep communications concise and understandable, particularly during complex incidents or when multiple incidents are in progress. The "Location-Command" designation should not change during the incident.

When the first FGC arrives on a fire rig, that unit loses its regular designation until its crew is moved up or until the original officer returns and places it in service.

The standard radio designation for the FGC is "Command" and automatically transfers with the current FGC throughout the incident. Anyone wishing to talk to the incident commander can simply request "Command."

The initial announcement of command signifies the beginning of any operation by a definite act. This announcement ensures quick, up-front command and also requires the FGC to commit a conscious act (per-

sonally) and a standard act (organizationally) by formally advising that he is the FGC. Now, everyone else en route, arriving, or operating at the scene knows that an FGC is in place. If no one has announced the assumption of command, the entire system knows that no one has begun FGC functions.

Those arrivng at an incident with an in-place FGC will fall under one of the following three categories:

1. Working under the FGC's command
2. Taking command, if passed by first FGC, or
3. Assuming command from FGC by virtue of higher rank.

COMMAND POSITIONING

THE COMMAND POST

The standard command position for the Fireground Commander will be a stationary one inside of a command vehicle or a piece of fire apparatus which is then called the "command post." It should be situated in a predictable and conspicuous location which affords the FGC a good view of the fire building and surrounding area. When possible, it should be in front of the fire and should NEVER interfere with apparatus movement. Ideally, it would also offer a view of two sides of the fire building.

The entire fire command system revolves around the rapid establishment of this command post. Utilizing long-term, mobile command or multiple command posts defeats the operation. In the cases where a company-level officer establishes a mobile command, he should upgrade to a stationary command as soon as possible using the standard procedures.

The FGC must discipline himself to remain àt the command post and manage the incident from one basic position. The system is designed to support and assist the centrally located FGC. A consistent command post also eliminates the too-common question, "Where's the Chief?" Now, everyone on the fireground knows where the FGC is, how to contact him by radio, and generally what he is doing.

Advantages of The Command Post

When a standard position inside a particular vehicle is assumed by the FGC, it becomes his *"field office"* and gives him the following advantages:

- Stationary position
- A quiet place in which to think and decide
- A vantage point from which to see
- Inside lighting
- A place to write and record
- More powerful radios
- Reference and preplanning material
- Protection from the weather
- Space for staff
- Computers (in some systems).

FIGURE 3.1.3: Very mobile commanders must adjust to a fixed command post.

Very mobile commanders may require a period of adjustment to the new experience of being tethered to the command post, particularly after their rough-and-ready days of moving all over the fireground to look, check, order, and verify in person. Everyone in the "new and improved" system should support and assist his efforts to "unlearn" the old habit of orbiting the fireground. Improved command and control experiences will also reinforce the stationary FGC approach.

SUMMARY

An effective overall management system is required at the fireground. Operations at the task level require a strategic plan, application of SOPs, and continual management.

A command procedure is necessary to have everyone cooperate to fix command on one individual, ensure strong, direct, and visible command as quickly as possible, and establish a management system that will provide clear, concise objectives and define functional relationships.

For fireground operations to be successful, there must be agreement that the FGC will act as the overall manager and everyone else will follow and support his plan.

The command post offers the continuous ability to communicate effectively. Since radio contact is the major link to all units, it must be available at all times.

When the FGC leaves the stationary position, he necessarily loses all its advantages including the use of his staff personnel. He must learn to stay put and manage the system from the command post.

In most cases, the first-arriving unit or officer should act as the FGC until relieved by a higher ranking officer, or until the incident is terminated. The assumption of command at this point should be mandatory. The first-in officer should decide on commitments based on nothing showing mode, fast attack mode, or command mode.

Decisions made by the first-in officer are greatly influenced by the arrival time and sequence of responding units.

The Company Officer in the normal command position can "move up" an acting officer, assign company personnel to aide functions, or assign company personnel to another officer.

The person assuming command must confirm his command. He should inform his central communications by broadcasting his unit designation, arrival, assumption of command, and the name and location of his command post. He is to be designated "Command."

Those arriving at an incident having a FGC will work under his command, take command if it is passed on, or assume command (if they are of higher rank using standard transfer procedures).

As soon as possible, a standard command post should be established in a stationary position. the FGC, with few exceptions, should remain at the command post throughout the entire operation. One of the major advantages to this command post is the effective communications capability it can provide.

COMMAND DEVELOPMENT

The FGC must work to develop standard procedures for the assumption, confirmation, and effectiveness of command. You should know and practice how to assume initial command, transfer command, and take over a mid-incident command.

The command post is a critical part of central command. As a FGC, you should know how to use its advantages and equipment to the maximum effectiveness. This will require you to keep up to date with the latest equipment and procedures.

The following report card is provided so that you can evaluate your FGC knowledge and skills in classroom exercises, simulations and on the fire scene.

Fireground Commander Report Card

Subject: Assumption, Confirmation, and Positioning of Command

Did the Fireground Commander:

☐ Correctly assume command?
☐ Confirm his command?
☐ Select the proper command mode?
☐ Set up a command post as soon as possible?
☐ Correctly turn over command, accept a mid-point command, or continue command throughout the incident as required?

3 FUNCTIONS OF COMMAND

Section 2
THE SITUATION EVALUATION

MAJOR GOAL

TO DEVELOP A REGULAR APPROACH TO SITUATION EVALUATION USING THE STANDARD FORMS OF INFORMATION AND FIREGROUND FACTORS.

OBJECTIVES By the end of this section, you should be able to:

1. Define "size-up." (p. 38)
2. List five exterior conditions that are easily identifiable through visual observation. (p. 39)
3. List the major advantages offered by a prefire plan. (p. 40)
4. State the three fire problems of occupancy preplanning. (p. 41)
5. List and describe the four weight classes for building evaluation. (p. 41)
6. List four questions that should be answered by the prefire plan. (p. 42)
7. Correctly identify the prefire planning symbols. (p. 43)
8. List the five major categories recorded on the tactical worksheet. (p. 44)
9. In your own words, describe the advantages of the tactical worksheet. (p. 44)
10. Match corresponding actions to the standard stages of a structure fire. (p. 46)
11. Define "fireground factors." (p. 47)
12. List the eight major categories of fireground factors. (p. 47
13. Explain the "handful of factors" rule. (p. 47)
14. List and define the three factors affecting information management. (p. 49)
15. Define and give examples of fixed fireground factors. (p. 50)
16. Define and give examples of variable fireground factors. (p. 50)

SITUATION EVALUATION

THE SIZE-UP

The second basic fireground management function is the situation evaluation. The initial phase is known as "size-up" and occurs at the beginning of fireground operations. Size-up is a systematic process consisting of the rapid, yet deliberate, consideration of all critical fireground factors and leads to the development of a rational attack plan based on these factors. The size-up cannot be delayed, nor can it be a time-consuming process.

This initial evaluation produces the information the FGC must have to make decisions and take action. Unfortunately, the process usually begins at the worst time on the fireground, when all the needed facts are difficult, if not impossible, to gather. The process of situation evaluation remains difficult throughout the early stages of the operation as the FGC attempts to gather facts and verify their completeness and accuracy within a very compressed time frame.

It is not uncommon for fire departments to begin operations before adequately reviewing all of the critical fireground factors. Fire attack can be an instinctive and action-oriented process that involves taking the shortest and quickest route directly to the fire. Action feels good on the fireground; thinking delays action. Beware of a nonthinking attack and nonthinking attackers.

The initial evaluation is the basis of the attack plan and is extremely important to overall success. Quick, correct action at the beginning will often allow the FGC to structure an effective primary search and fire control operation.

It is an unfortunate fact of life that the fireground is a difficult setting for management. Everyone is excited, it is tough to communicate (no one can hear so everyone shouts), and most firefighters tend to have their own attack plans. The FGC must often begin operations with inaccurate and incomplete information—truly a manager's nightmare.

One of the biggest problems for the FGC is the constant consideration of almost endless amounts of data. He must become an information processor; receiving information on conditions, translating it into tactical decisions, and then ordering companies to perform operations to implement his decisions. Obviously, the FGC must develop a simple, efficient system to deal with fireground information. His effectiveness increases directly with his ability to standardize the situation evaluation.

The Initial Evaluation

The FGC must place himself in the strongest possible position for a fast, initial evaluation. This requires a certain amount of talent. He must be able to consistently obtain good fireground information and translate it into a usable form; the information is used to develop a plan structured around the tactical priorities of rescue, fire control, and property conservation.

The initial evaluation begins at the time the alarm is received. Dispatch can provide useful information, including the type of call, occupancy, general area, and the units responding. While en route, the FGC can observe weather conditions, note the time of day and receive additional information such as reports of persons trapped. The

FGC considers all of this information, beginning the situation evaluation before arriving at the scene.

As the FGC approaches the scene, he can add any visible signs to his data base along with an initial impression of fire conditions. Therefore, how he approaches the scene can be very important. When possible, the approach should take a route that partially or completely circles the fire. This "drive by" system may delay his official arrival by a few seconds, but it may well provide him with significant facts that are not visible from the command post. These facts may often apply directly to potential structural failure or rescue problems.

Visual Observation

Proper evaluation is greatly influenced by the location of the command post. When the FGC selects a location for his command post, he must keep in mind the view it will provide. From the command post, the FGC must be able to observe the general effects of firefighting actions. This means placing the command vehicle close enough to see the operation, yet far enough back to afford a wide angle view of the fire area.

There should be no middleman in this process. Visual observation is a personal, rapid method to size-up fire conditions, and the FGC should keep an experienced eye on the fire throughout the operation. The FGC has to realize that certain conditions are often visible outside of a structure before they are noticeable inside.

FIGURE 3.2.1: Visual observation shows fire conditions and firefighter action.

The FGC's visual observations may be the best source for information on exterior conditions. Some of the exterior conditions that are easily identifiable visually include:

Area arrangement—streets, buildings, potential exposures, access obstacles

Fire building detail—building type, size, height, occupancy, construction, age and general condition, and structural stability
Fire conditions—what's burning, size, location, products of combustion
Resource status—Placement and use of apparatus and activities of personnel
Effects of firefighting action—basic operational process.

Looking out from the command post is an essential data gathering method, but it is limited to the field of vision available from that one spot. This handicap can be balanced by the use of prefire plan data and reconnaissance reports from the different sectors on the fireground. Obviously, the FGC will have to rely more heavily on progress reports from Interior Sectors during offensive interior attacks than during defensive "surround and drown" operations.

THE PREFIRE PLAN

A major support item during the evaluation phase is the prefire plan. Most fire departments perform these planning activities as part of a structured program that includes touring significant occupancies. These tours identify and record any important characteristics that would effect firefighting. The information is noted and usually stored in notebooks which are typically carried in apparatus or command vehicles.

Prefire planning is conducted with the advantage of ideal conditions. Tactical surveys are carried out in daylight under nonfire conditions. There is no sense of urgency. Firefighters have time to visit, decide, contemplate, draw, and even redraw until they get it right.

Conversely, conditions during the fire bring about the opposite reactions. The FGC and his firefighters must function in front of and inside a burning building, often at night, with little time to make detailed observations. Under these conditions, it is equally difficult to decipher elaborate preplans. The FGC will usually have only one chance to make the right decisions, in a very short period of time. Realistically, this means that prefire plans are drawn under the best conditions but generally used under the worst conditions. This adds greater emphasis to prefire tours.

These information-gathering safaris increase awareness and familiarity for the firefighters who may have to operate at that location under fire conditions. Such preresponse homework is essential to the safety of personnel and the overall fireground operation. Even though the attack units may not usually refer to the plan during a fire, Command Officers are in a position to use the plans and can relay pertinent information to the forward sectors and companies.

Ongoing training activities should include the use and revision of these preplans. The results will be an increased familiarity with the specific occupancies and up-to-date plans. When the plan is written only to be buried in a binder, it will not be effective when needed.

REMEMBER: Prefire planning arms the FGC with facts that are impossible to acquire under fire conditions.

FIGURE 3.2.2: Prefire tours allow information to be gathered under the best of conditions.

Occupancy Preplanning

A starting point for occupancy preplanning is an evaluation of the fire problems associated with the building in terms of size, hazard, and built-in protection.

SIZE—The size of the building and the potential size of the fire provides a description of the scale of possible operations.

HAZARD—The analysis of hazards should include the number, location, condition, and activity of occupants; the amount, nature, location, and arrangement of fire load; assess characteristics that could obstruct normal operations; present or potential water problems; possible delayed or inadequate response; and any special hazards associated with the structure or the area.

BUILT-IN PROTECTION—The benefits provided by automatic extinguishment, systems early warning devices, adequate separation, fire-resistive construction, and firefighting support equipment must also be considered.

The size and hazard factors add to the risk, while built-in protection subtracts from it. The final evaluation combines all of these factors to produce an *overall building risk rating*. The rating can be expressed in terms of "weight" classes:

Super weight—buildings that present huge life-safety and fire problems that would require a major fire fight, involving all area resources for an extended period of time and serious potential danger to personnel

Heavy weight—large buildings that present significant fire and rescue problems requiring a greater alarm response for control

Middle weight—medium-sized buildings that present a low level fire problem requiring routine first-alarm tactics

Light weight—single family and small commercial risks controlled by single attack teams. They present a low fire problem with predictable outcomes.

Prefire Plan Considerations

Since the purpose of a prefire plan is to provide the FGC with information on the critical factors he cannot see from the command post, a prefire plan should answer the following questions:

1. What factors are present?
2. What does the FGC need to know to be effective?
3. What factors can be seen from the command post?
4. How serious a problem can be caused by the unseen hazards?

Answering these questions at the beginning of the preplan process may change the selection of the buildings that are preplanned. For example, if a building's only significant feature is its size, it may be preempted for a smaller building with major internal hazards or features that are not identifiable from the exterior. Knowing these details will help the FGC operate more safely and effectively.

Prefire Plan Management

Filing, storing, and finding preplans can present many practical and functional problems. Command must be able to quickly access the correct plan. If he can't find it, he can't use it.

These plans are usually best stored in notebooks that are kept in apparatus cabs or car trunks. For quick access, space should be provided for storage. Given the reality of physically manipulating binders while inside a vehicle, it is best to carry the preplans for the first alarm response area only. The fewer books carried, the better their usage.

When used correctly, prefire plans are employed at the beginning of the suppression effort, when time is critical. A simple filing system using occupancy/address will help the FGC to find the correct plan in the book. Some departments operate with microfiche or computer-aided text and graphic systems using video terminals to access and display the data.

The Prefire Plan Format

The preplan format is important if the plan is to be a regular component of fireground operations. If the plan is too complicated or too difficult to deal with under fire conditions, it will not be used. The plan layout should present primary information using a graphic- and symbol-oriented approach. The primary data should include physical layout, features to be used for firefighting, and hazards to firefighting personnel. Particular emphasis should be placed on potential problems and obstructions to access.

At all costs, the format should avoid excessive detail. This distraction affects the most critical factor, time. Using a pure scale drawing, some of the more important features may be too small to be readily identified. The graphic presentation should provide for these critical factors to stand out, yet not get lost in details.

The preplan should direct attention to the features which will affect tactical decisions and firefighting. Drawings should assume that the FGC will be positioned in front of the building or in an alternate predictable location and coincide with the view and orientation he would have. Standard fire protection symbols on diagrams present an easy way to locate major factors that will affect operations. Unfortunately, some of the typical symbols devised for fire engineering risk evaluation are highly technical and tend to hide crucial features.

Figure 3.2.3 shows *prefire planning symbols* that have been designed specifically for that purpose. They create an instant focus on the critical factors, improving the chances for a quick, correct response.

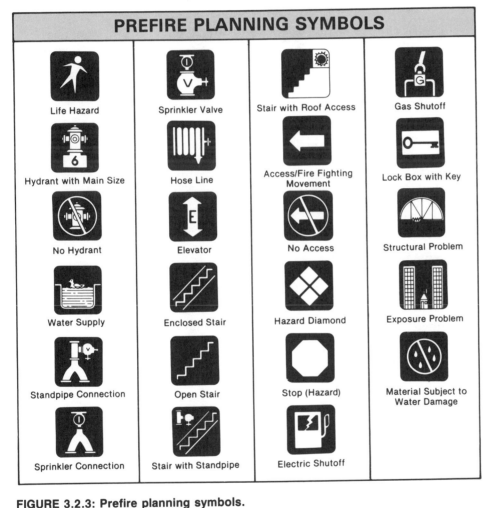

PREFIRE PLANNING SYMBOLS

Life Hazard	Sprinkler Valve	Stair with Roof Access	Gas Shutoff
Hydrant with Main Size	Hose Line	Access/Fire Fighting Movement	Lock Box with Key
No Hydrant	Elevator	No Access	Structural Problem
Water Supply	Enclosed Stair	Hazard Diamond	Exposure Problem
Standpipe Connection	Open Stair	Stop (Hazard)	Material Subject to Water Damage
Sprinkler Connection	Stair with Standpipe	Electric Shutoff	

FIGURE 3.2.3: Prefire planning symbols.

THE TACTICAL WORKSHEET

The FGC must keep track of the various task assignments and the organizational development he has created on the fireground. This can prove to be a significant problem as operations become more complex or are forced to change rapidly. He needs a system which allows him to write and record all important firefighting activities. The *tactical worksheet,* designed to be used by the FGC at the command post, provides a standard system that is understood by all the players.

The tactical worksheet relates fireground action in the same way that prefire plans record information before the fire. These forms go together in a practical way to integrate information management, fire evaluation, and decision making. The use of both forms, often together, create a routine management approach to every fire.

The tactical worksheet must be easy to understand and complete if it is to be used on the scene. The well-designed worksheet will systematically lead the FGC through considerations and decisions, acting as a "memory jogger." To be easy to use and effective, the form should utilize a check the box, fill in the blanks, and complete the drawing approach. It should record the following data:

- Unit status—responding, staged, assigned
- Benchmark of completion and times
- Diagram of fire area or occupancy (including highrise floors)
- Activity check list—sectors, rescue, fire control, property conservation, safety, and special operations
- Organization structure.

Using the tactical worksheet should be a normal part of the FGC's routine, beginning upon his arrival. Recording the details of assignments, actions, and locations as they occur eliminates having to play catch up. It is much easier to record events as they happen than it is to reconstruct them after the fact. Usually, any attempts to locate and fill in the blanks on a command status board in the middle of an action will fail. The process of recording orders and actions as they occur will strengthen the FGC's conscious, deliberate, decision making. By using this worksheet, he will build a set of well organized, functional management habits.

Tactical Worksheet Advantages

The tactical worksheet offers many advantages to the FGC. It:

- Matches and supports the regular management system
- Records information in a standard place
- Supplies reminders of key points
- Provides the same form for everyone
- Standardizes communications and information management
- Improves understanding at the command post
- Supports the command transfer process
- Compiles useful information for the command post staff
- Furnishes an ongoing tactical accounting and inventory system.

The worksheet allows the FGC to establish control at the beginning of the operation, when control is the easiest to capture. It will also increase his ability to maintain control later, when firefighters are operating in more places, doing more things. This detailed awareness of operations is well worth the time invested completing the worksheet. The best way to avoid confusion is to maintain control, never letting confusion surface. No one else will maintain control for the FGC. This is one responsibility that always rests with the big buckaroo (FGC).

THE EVALUATION SYSTEM

A functional fireground evaluation system considers current conditions to create quick action and forecasts outcomes to prevent sur-

TACTICAL WORK SHEET

INCIDENT NO. _____   TIME

☐ 2 + 1   ☐ FIRST ALARM   ☐ _____

ADDRESS: _____

OCCUPANCY: _____

☐1 ☐2 ☐3 ☐4 ☐5 ☐6 ☐7 ☐

INITIAL REPORT:
☐ INITIAL REPORT:
☐ CMD LOCATION
☐ ALL CLEAR
☐ STANDPIPE
☐ SPRINKLER
☐ INVESTIGATOR
☐ PUMPED WATER
☐ P.D.
☐ GAS   ☐   ELECT.
☐ LEVEL 2 STAGING LOCATION

BY _____
☐N ☐S ☐A
☐E ☐W ☐B
☐

☐ UNDER CONTROL _____

STAGED _____

E		
E		
E		
E		
L		
L		
H		
R		
U		
BC		

SECOND ALARM

E		
E		
E		
E		
L		
L		
H		
R		
U		
BC		

QUADRANT & EXPOSURES

```
        3
  2   B | C   4
      A | D
        1
```

COMMAND

FIGURE 3.2.4: A tactical worksheet.

prises. A fast, initial size-up gives the FGC a "snapshot" of the existing conditions at the beginning of operations. He must, then, evaluate the critical factors, develop a tactical plan, translate the plan into tasks, and then assign companies to work on those tasks.

The FGC places himself at a serious disadvantage when he considers conditions only within the single dimension of current time. He will constantly be surprised by changing conditions. Sometimes the fire will outrun the attack plan. In order to stay ahead of the fire, the FGC must use an evaluation system that considers and accounts for changes. If he does not do this, the attack plan will be forever out of time or out of position.

This time/factor evaluation considers and combines time and standard conditions. Time moves forward in a predictable way, and standard conditions keep changing in predictable ways. The matching of times and conditions produce a scale that represents what really occurs on the fireground. Standard evaluation allows the FGC to visualize and operate through the fireground cycle, from beginning to end. The evaluation allows Command to pinpoint exactly where a specific fireground operation is on the scale. More importantly, it allows the FGC to project his evaluation along the scale and, essentially, predict the future.

Matching Actions To Conditions

Structure fires progress through fairly standard stages. Actions can be matched to fire conditions on a one-to-one scale. Figure 3.2.5 shows this correlation. This scale gives the FGC a picture of the full range of fire stages and begins to link the proper firefighting actions and resources to conditions with time progressions. The scale provides an operations curve which outlines the beginning, middle, and end of the fire and the link between them.

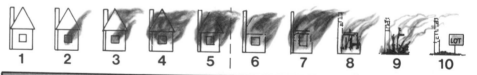

FIRE STAGES AND CORRESPONDING ACTIONS

1. Nothing showing	1. Investigate
2. Smoke showing	2. Fast, aggressive interior attack
3. Smoke and minor fire showing	3. Fast, aggressive interior attack
4. Working fire	4. Well contained interior attack
5. Deep-seated working fire	5. Cautious, interior attack
6. Marginal working fire (*offensive/defensive*)	6. Cautious, interior attack with preparations for exterior, defensive operation
7. Fully involved	7. Exterior, defensive operation
8. Coming down	8. Exterior operations which anticipate structural collapse
9. Down	9. Exterior operations which anticipate structural collapse
10. Parking lot	10. Remove preplan from file

FIGURE 3.2.5: Actions can be matched to standard fire stages.

FIREGROUND FACTORS

DEFINING FIREGROUND FACTORS

The FGC needs a simple system which will allow him to deal with all fireground information. Without such a system, it will be impossible to perform the situation evaluation and all other command functions efficiently. Fireground factors offer such a structure.

Fireground factors are a list of the basic items that the FGC must consider when evaluating tactical situations. They provide a checklist of the major topics involved in size-up, decision making, initiating operations, and review and revision.

Command deals with these factors through a systematic management process that sorts out the critical factors in priority order and then seeks more information about each of them. These factors are:

- Building
- Fire
- Occupancy
- Life-Hazard
- Arrangement
- Resources
- Action
- Special Circumstances.

The FGC must keep himself informed on the fireground. New decisions must be made and old decisions revised based upon increased data and improved information. He can never assume that the action-oriented fireground actors engaged in firefighting activities will naturally feed him a continuous supply of top-grade information.

The FGC must develop a standard approach to sorting out and prioritizing factors. He must train himself to continually engage in conscious information management. Fireground factors offer the basis for a *standard information management approach.* Decisions can be no better than the information on which they are based. A standard approach is the key to fireground performance and must become habitual.

During most critical fire scenarios, Command must develop an initial attack plan based on incomplete factor evaluation. Even though he will continue to improve its quality, he will seldom function with complete or totally accurate information on all the factors. This is most evident during initial operations.

INFORMATION MANAGEMENT

Any manager can only deal with a limited number of factors of any kind, at any one time. He cannot make an effective decision from 150 pieces of information—he can deal with 5 or 6. The inclination to deal with too many factors will soon overload the FGC, resulting in chaos and confusion. Considering this *"handful of factors"* rule, the identification of the critical factors now becomes crucial. Fortunately, all factors are not critical in any given tactical situation. Once the critical factors have been identified, Command must focus his time and energies on them. The major fireground factors are listed in Figure 3.2.6. If all were critical for all times during all fires, the FGC's information management task would be impossible.

FIREGROUND FACTORS

BUILDING
- Size—area and height
- Interior arrangement/access (stairs, halls, elevators)
- Construction type—ability to resist fire effect
- Age
- Condition—faults/weaknesses
- Value
- Compartmentation/separation
- Vertical-horizontal openings, shafts, channels
- Outside openings—doors and windows/degree of security
- Utility characteristics (hazards/controls)
- Concealed spaces/attack characteristics
- Exterior access
- Effect the fire has had on the structure (at this point)
- Time projection on continuing fire effect on building
- How much of the building is left to burn?

FIRE
- Size
- Extent (percent of structure involved)
- Location
- Stage (inception ———→ flashover)
- Direction of travel (most dangerous)
- Avenue of travel
- Time of involvement
- Type and amount of material involved—structure/interior/finish/contents/everything
- Type and amount of material left to burn
- Product of combustion liberation (smoke, heat, flame, gas, etc.)
- What is perimeter of fire?
- How widespread is the fire area?
- Fire access—ability to operate *directly* on fire

OCCUPANCY
- Specific occupancy
- Type-group (business, mercantile, public assembly, institutional, hazardous, industrial, storage, school)
- Value characteristics associated with occupancy
- Fire load (size, nature)
- Status (open, closed, occupied, vacant, abandoned, under construction)
- Occupancy—associated characteristics/hazards
- Type of contents (based on occupancy)
- Time—as it affects occupancy use
- Property conservation profile/susceptibility of contents to damage/need for salvage
- Moral hazard

LIFE HAZARD
- Location of occupants (in relation to the fire)
- Number of occupants
- Condition of occupants (by virtue of fire exposure)
- Incapacities of occupants
- Commitment required for search and rescue (men, equipment, and command)
- Fire control required for search and rescue
- EMS needs
- Time estimate of fire effect on victims
- Exposure/control of spectators
- Hazards to fire personnel
- Access rescue forces have to victims
- Characteristics of escape routes/avenues of escape (type, safety, fire conditions, etc.)

ARRANGEMENT
- Access, arrangement, and distance of external exposures
- Combustibility of exposures
- Access, arrangement and nature of internal exposures
- Severity and urgency of exposures (fire effect)
- Value of exposures
- Most dangerous direction—avenue of spread
- Time estimate of fire effect on exposures (internal and external)
- Barriers or obstruction to operations
- Capability/limitations on apparatus movement and use
- Multiple buildings

RESOURCES
- Manpower and equipment on scene
- Manpower and equipment responding
- Manpower and equipment available in reserve
- Estimate of response time for men and equipment
- Condition of men and equipment
- Capability and willingness of personnel
- Capability of commanders
- Nature of command systems available to command
- Number and location of hydrants
- Supplemental water sources
- Adequacy of water supply
- Built-in private fire protection (sprinkler, standpipe, alarms)

ACTION
- Effect current action is having
- Things that need to be done
- Stage of operation (rescue, fire control, property conservation)
- Effect of the command function—is command established and working?
- Is there an effective organization?
- Has FGC forecast effectively?
- Is there an effective plan?
- Tactical priority questions: Are victims okay? Is fire out? Is loss stopped?
- What is the worst thing that can happen?
- Are operating positions effective?
- Are operating functions effective?
- Are there enough resources? (Personnel, apparatus/equipment, logistics/support, command, water, SCBA air)
- Are troops operating safely? Do you fear for their lives?
- Situation status:
  Continuum   Under control   Out of control

SPECIAL CIRCUMSTANCES
- Time of day/night
- Day of week
- Season
- Special hazards by virtue of holidays and special events
- Weather (wind, rain, heat, cold, humidity, visibility)

FIGURE 3.2.6: Fireground Factors.

Dealing with fireground information becomes a complex problem in most tactical operations. Some factors can be observed directly from the command post, while others can only be determined from different locations inside or outside of the structure. Obtaining fireground factor information requires the FGC to develop and evolve a system of fireground intelligence.

Information management revolves around three basic factors:

1. **VISUAL**—information obvious to visual inspection. It is usually acquired by looking at the situation from the outside, involving the critical, perceptive capabilities of the FGC. This is the most common and natural factor.
2. **RECONNAISSANCE**—information not available visually to the FGC at the command post. It is acquired by assigning personnel to specific problems and locations and receiving their information-oriented reports.
3. **PREFIRE PLANNING AND FAMILIARITY**—information gained from formal prefire planning and informal familiarization activities. This increases information beyond what is quickly available from the outside and provides intelligence not normally available or requiring reconnaissance assignments to obtain.

There is a strong temptation for the FGC to depend too greatly on the visual factor and try to view conditions by physically wandering all over the fireground. This walking survey can and usually does produce a serious breakdown in the command structure. No one can command rationally while on a dead run.

Complex scenarios present widespread and dynamic (always changing) settings which absolutely defy the capacity for on-site appraisal by one person. Decentralized units and sectors must be used as information centers. No amount of roadwork will enable you to run fast enough or far enough to keep yourself informed.

Reporting should be combined into regular assignments over the entire fireground. It must be an internal, continuous reponsibility of every unit. Important information can be funneled upward through the sectors which comprise the FGC's management network.

This approach requires a simple, yet standard communications SOP for the entire organization. During difficult fireground operations, everything works against communications; therefore, personnel must be trained and disciplined to communicate effectively.

Fireground factors are the basis for the prefire plans which should include only the critical factors relating to that occupancy. Prefire plans should be created with the realization that their principle purpose is to provide critical information under difficult conditions: they should be simple, easy to read, and focused on specific data.

Fireground Factor Management

Effective factor management is an ongoing part of operations which must be integrated into all activities. The absence of one or more *critical factors* cannot become a distraction. The FGC must deal or compensate for the lack of the information he does not have, in relation to what he does have. Often, what information is received becomes a clue to what additional information needs to be persued.

Information concerning fireground factors will arrive at different times. The FGC learns some things before the fire, some during, and some after. The challenge is to quickly acquire useful information and to use it effectively.

The fireground factors represent an array of items that are dynamic throughout the entire process. Accordingly, the relative importance of each changes over time. Command must continuously deal with these changes and base decisions on current information. The effective FGC does not stick with his initial plan of attack after conditions have changed. Fire operations require him to continually reconsider these factors based on data feedback and to make attack plan revisions.

The FGC must concentrate not only on the right factors, but also on the proper time to ask for information and decide on actions—the overall action plan. The action plan's tactical priorities dictate the basic order of fireground operations. Command must identify the best source of information for a particular problem at any given time. He must keep in mind that the source for some information may change with time.

Some fireground factors are fixed and rigid; some are variable and flexible. All of the factors influence the fire and must be considered, but the FGC's ability to change any of them is limited. For example, he cannot change the weather, the occupancy, the type of construction, or the size of the building (although it will get smaller as the fire gets older).

He must then concentrate on the factors that he can alter, manipulate, remove and create. The question at this point becomes, ''What factors can I change?'' The answer is integral to the plan of attack. Fixed factors lend themselves to prefire planning activities while the variable factors are best managed by using visual and reconnaissance methods while the fire is in progress.

Virtually all factors exist on a scale or spectrum; they are not absolute. Size-up involves first selecting the critical factors and then evaluating their severity. This severity can range from ''no problem'' to ''absolutely critical.''

Since factors are not static, they generally get worse if left alone (although eventually they all will go away). Usually the FGC develops decisions and initiates actions in response to the critical factors. Once the critical factors are identified, reducing the severity of these factors becomes the major activity on the fireground.

SUMMARY

The second basic FGC management function is the situation evaluation. This requires a size-up—the rapid, yet deliberate consideration of critical fireground factors and the development of a rational attack plan based on these factors. Command effectiveness increases directly with the FGC's ability to standardize this evaluation.

The initial evaluation begins with information received from Alarm at the time of the alarm, continues with information received while in route, improves with information collected during the "drive-by," and turns into a rapid process once on-scene visual evaluation begins.

There are no middlemen in the visual evaluation process. Exterior conditions easily identifiable by the FGC include overall area arrangement, basic fire building details, fire conditions, resource status, and the effect of firefighting action. Preplan data and reconnaissance reports add to the data bank.

The prefire plan can be critical to the evaluation phase. All prefire tours should note and record any important characteristics that would affect fire fighting. Prefire planning should review the fire problems of the building in terms of size, hazards, and built-in protection.

Analysis of the fire problems produces an overall building rating in terms of weight classes (super weight, heavy weight, middle weight, and light weight).

The prefire plan should tell the FGC what factors are present, what factors he must know to be effective, what factors can be seen from the command post, and how serious the unseen hazards are.

To be effective, the prefire plans must be organized, easy to locate, formatted to emphasize important factors, be easy to read, and utilize standardized prefire planning symbols.

The tactical worksheet will assist the FGC in keeping track of the various task assignments and organization development on the fireground. It can be used with the prefire plan to provide a practical way to integrate information management, fire evaluation, and decision making.

A tactical worksheet should record unit status, benchmark times, a diagram of fire area and occupancy, an activity checklist, and a description of the organization structure. A well-designed worksheet will match and support the regular management system, record information in a standard place, supply reminders of key points, standardize communications and information management, establish a method of command transfer, and furnish an ongoing tactical accounting and inventory system.

The FGC needs to do a time/factor evaluation. This requires matching specific fireground actions to the standard stages of a structure fire.

Fireground factor analysis offers a simple, effective system to deal with fireground information. These factors provide a list of major items that must be considered when evaluating tactical situations. The major factor categories are building, fire, occupancy, life hazard, arrangement, resources, action, and special considerations.

Factors will change with time. The FGC must establish an organization that will provide him with needed information. He must use a standard approach to sort out factors and prioritize them. He should limit the number of factors being considered to avoid factor overload.

The FGC must be prepared to deal with information from visual, reconnaissance, and prefire plan sources. The FGC should manage factors as an ongoing process, keeping the process dynamic. He must know what factors to consider and when they are important. Effective management calls for the realization that some factors are fixed and rigid, while others are variable and flexible. He should concentrate on factors he can change.

The FGC develops decisions in response to critical factors. His task is to reduce the severity of these factors through fireground activities.

COMMAND DEVELOPMENT

An effective FGC must be able to carry out a situation evaluation. This requires you to obtain information from Alarm and personnel at the scene, and to use your own visual observations. To complete a successful evaluation, you MUST be able to perform a rapid, systematic size-up that is based on the critical fireground factors that apply specifically to the incident.

Practice making size-ups of firegrounds and simulations, noting area arrangement, building detail, fire conditions, resource status, and the effects of firefighting action. Use the prefire plan to assist with your evaluation.

For all fires and simulations, use a tactical worksheet to help improve your management skills. Keep an account of fire stages and corresponding actions. Learn to use critiques to help you assess the actions taken.

The following report card is provided so that you can evaluate your FGC knowledge and skills in classroom exercises, simulations, and on the fire scene.

Fireground Commander Report Card

Subject: Situation Evaluation

Did the Fireground Commander:

- ☐ Conduct a rapid, systematic size-up?
- ☐ Utilize information from Alarm?
- ☐ Gain information from visual observations?
  - ☐ Area arrangement
  - ☐ Building details
  - ☐ Fire conditions
  - ☐ Resource status
  - ☐ Effects of firefighting action
- ☐ Use the prefire plan?
- ☐ Identify the critical fireground factors?
- ☐ Keep an accurate tactical worksheet?
- ☐ Manage variable factors?

3 FUNCTIONS OF COMMAND

Section 3
COMMUNICATIONS

MAJOR GOAL

TO INITIATE, MAINTAIN, AND CONTROL EFFICIENT FIREGROUND COMMUNICATIONS.

OBJECTIVES By the end of this section, you should be able to:

1. State the communications responsibilities of the FGC. (p. 54)
2. List the five major fireground communications problems. (p. 54)
3. State at least two reasons why the FGC must initiate communications upon arrival at the scene. (p. 56)
4. List what the initial report should include. (p. 57)
5. List three elements of a sector report. (p. 58)
6. List four elements of a fire company report. (p. 59)
7. Describe the role of the command post staff in regard to communications. (p. 59)
8. State which factor is the controlling factor in communications. (p. 59)
9. List the standard communications guidelines. (p. 60)

COMMUNICATIONS

INTRODUCTION

The third basic fireground command function is initiating, maintaining, and controlling the communications process. Communications provide the connection between management (the FGC) and personnel (the working units), as well as the FCC's link to the outside world (Alarm).

Each level of the fireground organization has a somewhat different need and capability to communicate. Those differences will necessarily affect the entire communications process as fire operations continue. The FGC is on the command level and deals with decision-making, assignments, coordination, revision, and control as he determines the overall strategy and manages the attack plan. His position is stationary and remote in the command post, where he uses a mobile radio as his major communications tool. Of all the participants, the FGC is in the best position to communicate. He also has the highest need to communicate and depends most on this process to do his job.

COMMUNICATIONS PROBLEMS

Communications problems are considered the most common operational snag in the majority of departments, effecting the firefighters' ability to start, coordinate, and complete effective operations.

Predictable fireground communications problems include:

LACK OF SOPs—A communications game plan is an essential part of the entire fireground SOP package. It provides a uniform approach for everyone in the system.

DEFICIENCIES IN TRAINING—Even though most of the fire actors use the radio on a daily basis, there is still the need for a basic "push-to-talk" training program. The fireground is a difficult place for effective communications. Noise, excitement, and radio volume are usually high, while radio discipline and control are often low.

ORGANIZATION PROBLEMS—Communications often becomes the "fall guy" for organizational problems. Multiple commanders, no commander, lack of an attack plan, and general confusion can overwhelm the best communications system.

EQUIPMENT PROBLEMS—Hardware problems can, and do, effect the entire operation. Poor reception or insufficient channels usually cannot be corrected on-scene. Effective communications results from a match of good system design and disciplined participants, not from fancy equipment.

COMMUNICATION TECHNIQUES—Personal techniques such as voice levels, word choice, timing, level of excitement, and degree of patience have a direct effect on the ability to communicate. Verbal cues may indicate sarcasm or reinforce understanding. Technique problems require straightforward diagnosis and corrective coaching, beginning by listening to radio transmissions and then instituting the appropriate training loop.

FORMS OF FIREGROUND COMMUNICATIONS

There are four basic forms of fireground communications. The FGC must use a combination of these approaches to maintain command effectiveness. They are:

FACE-TO-FACE—This is the best communications form because the participants can combine a variety of interpersonal methods. As they talk, they can look at one another to see how the other reacts, evaluating facial expressions, gestures, and body language. This form of communication is limited to the range of personal contact.

RADIO—Radio communications provide a remote capability when face-to-face communications are not possible. The advantages are speed and the ability to communicate over a large area. The main disadvantage is its one-dimensional characteristics—only voice. Fire operations require both a strong radio procedure and a plan among all the communications participants. The FGC may become remote from the action but may not be remote from control of this action.

COMPUTERS—Fireground computer systems are starting to show up in a few places. If you are in a department that does not use fireground computers, or you believe that your department is too small or too remote to use these devices, keep in mind the rapid advances made by the computer and telecommunications industries in the last 10 years. It is predictable that the fireground computer system will be a part of most urban and suburban departments and many rural departments by the end of this century.

These systems typically have main computers located at Alarm and mobile terminals in the field units. They can provide on-line dispatching and tactical information in hard copy or on a video display. This advanced approach gives the FGC the ultimate in information access and instant technical assistance. It also represents the maximum in electronic impersonalization—the computer never expresses anxiety or offers reassurance.

STANDARD OPERATING PROCEDURES—SOPs are not generally regarded as a communications form, but they provide the organization with the ongoing capacity to evaluate, and then solve, communications problems. They greatly reduce the difficulties in communicating routine actions. A critical analysis of fire operations will produce many opportunities to standardize and streamline the system. This approach creates a high level of predictability and confidence, eliminates a lot of routine traffic, and frees communication space and time for more critical traffic. SOPs can be a powerful and effective communications tool.

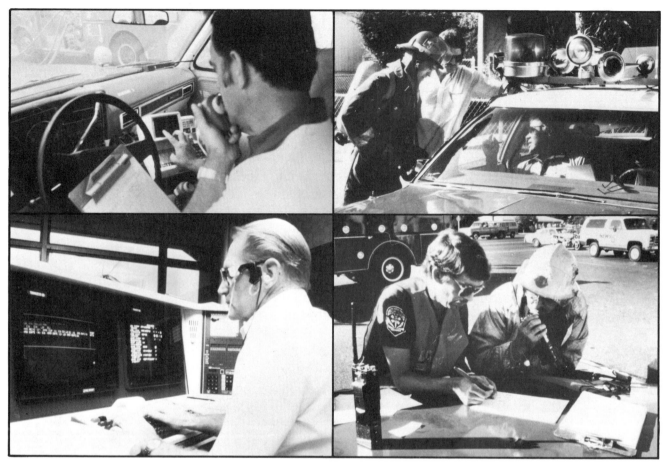

FIGURE 3.3.1: Effective operations require a combination of communication forms.

INITIATING COMMUNICATIONS

From the very beginning of fireground operations, the FGC must use communications to initiate and evaluate fireground actions. Upon arrival, he needs to advise all operating companies of the basic details of the attack plan. Throughout the operation, he must receive a steady feedback on the plan's effectiveness by way of reports from the Sector Officers. Therefore, it is essential that all operating companies and all Sector Officers have entered the communications loop once the FGC takes command.

THE INITIAL REPORT

The farther the FGC is removed from the action, the more he must depend on communications to enable him to fulfill his command functions and provide leadership. He begins with a brief initial report which explains the conditions he can see from the command post. This report is directed to everyone arriving at the scene or still responding to the scene, in addition to advising Alarm of the situation on arrival.

The initial report is not an affidavit of absolute accuracy. It is the best shot the FGC has at that moment to quickly develop and extend a description of what he can see. The report is updated as conditions change or additional information is received.

REMEMBER: Radio reports describe conditions from one position.

This initial report provides a standard description for the following items:

- Building size—small, medium, large
- Building height
- Occupancy
- Fire/smoke conditions—nothing showing, smoke showing, fire showing, working fire, fully involved
- Confirmation and designation of command
- Action being taken.

Since an organized beginning usually leads to a well-managed operation, the standard initial transmission becomes an indication of things to come, whether good or bad. For example, if Engine One is the first-arriving unit at North High School, the officer's radio report could be, "Engine One on the scene with a medium-sized, two story school with smoke showing. Engine One is laying a line in, making an attack on the north side. Engine One will be North High Command."

In this case, everyone responding knows that command has been established, what Engine Company One's crew is currently doing, and what type of building is involved. Responding firefighters can anticipate going into action on a working fire. At the same time, Alarm can begin to gear up the system to support a working fire incident.

This short transmission provides a description of conditions and announces the establishment of Command. It starts the process of FGC command and control at the very beginning. The initial fireground report helps the entire attack team start operations in an orderly, well-managed way, greatly increasing the chances for success.

The initial report also serves to initiate the communications process. The report sets the stage for two-way communications between the FGC and operating units and sectors.

MAINTAINING COMMUNICATIONS

THE FGC'S ROLE

The FGC gets the operation started by communicating. He decides on an overall strategy and attack plan and gives specific orders to companies and sectors to go to work. The emphasis in initial communications is on the FGC vocalizing what he wants accomplished. He is effective to the extent he can transmit clear, simple, understandable directions. Once the operation is set up, the FGC begins to receive feedback in the form of reports on the progress being made and the need for more resources or coordination. His success is dependent on this feedback.

As the operation goes on, the FGC must balance talking and listening and use the increased communications capability of the command

post to assist the crews. Critical listening is often the key element in effective fireground communications. (The system should be aware of those whose hearing is reduced by promotion.)

During the most active stages of fire operations, the FGC must use the physical and electronic advantages of the command post, making Command the ongoing focal point of fireground communications. This position allows him to initiate communications at the beginning of operations and maintain it during the active fire stages.

The FGC must understand that sectors and companies are on the most active, operational level of the fireground organization and realize how this affects their ability to communicate. The FGC must use his communications advantage to balance or overcome the disadvantages that go with the position and functions of the workers. He does this by taking advantage of the command post environment, being continuously available to respond (receive and transmit), and using his command post staff as communications helpers.

The ongoing availability of the FGC becomes an important factor in maintaining the communications process throughout the entire operation. From a practical standpoint, if an operating unit tries to contact the FGC and there is no answer, they will generally make one more attempt. If this second try fails, they will go back to work, choosing action and skipping communications. Their actions may prove to be outside the framework of the overall plan. This means that the FGC has lost some control of the operation. If action without communication goes on long enough, the free enterprise process takes over. The FGC has now lost control.

REMEMBER: Every FGC management function requires effective communications.

SECTOR REPORTING

During the initial stages of operations, the FGC will assign and directly supervise fire companies as quickly as they arrive. As the fire goes on and more companies arrive, he will begin to assign Sector Officers to various positions and functions. These officers become command partners, enabling the FGC to build an effective operation and to maintain a manageable span of control. They receive tactical assignments from the FGC and directly manage companies assigned to their specific sectors. Sector Officers use portable radios to communicate with the FGC and other operating and support sectors.

Sector Officers are usually near action and noise. They are generally not in a position to provide long dissertations. Their radio reports to Command should include:

- Positions
- Progress
- Needs (resources and support).

These three basic reporting items provide an adequate and simple information structure for the FGC to manage the attack plan and support sectors.

Fire Company Communications

Fire companies perform firefighting tasks, operating on the most action-oriented level. Their communications are even more basic than those of sectors and usually only need to include the following operational items:

- Assignment to a task with a location and objective (where and what)
- Resources and support required
- Supervision required
- Reassignment when the current task is completed.

Once companies have received their initial assignment, e.g., "open the roof," there is little need for additional transmissions unless they are faced with unexpected problems or barriers. Under active fire conditions, when operating companies do communicate, they generally have a critical need to do so.

Radio messages sent to fire companies must be task-oriented and indicate the objective. They should involve a function that fits the profile and capability of the receiver and can be reasonably handled by the company or groups of companies. Effective orders should tell the receiver where to go, to whom to report, and what to do. If the SOP system works, they already know how to do it.

Command must realize that the ability of firefighters to communicate is directly proportional to their relationship to the command post (attack teams using SCBAs will be harder to talk with than Sector Officers). This disadvantage can be overcome by the FGC making himself continuously available to respond to radio reports and updates.

As an operation grows in complexity, the FGC needs to delegate some of the responsibility for radio communications. Doing so does not free the FGC from his responsibility to maintain a constant awareness of what is going on. It does free him to listen and think. The FGC who attempts "Lone Ranger" management may find himself overwhelmed.

The capabilities of the entire team are improved by the use of the staff as communications technicians. A communications plan defining the use of multiple frequencies, radio procedures, and a team approach to command post operations is a major asset. Once the organization is in place, aides can communicate with support sectors (staging, resource, PIO, etc.) on separate frequencies, freeing up the FGC to concentrate on operating sectors. The whole command post staff should be directed toward equal parts of receiving, thinking, and transmitting.

CONTROLLING COMMUNICATIONS

THE FGC'S ROLE

In a very practical sense, the FGC's ability to control the communications process regulates his ability to command the operation. The fireground communications process is a mixture of equipment and standard procedures linked by human participants. Once the equipment and procedures are in place, the human factor becomes the deciding factor.

Since the FGC is located in a relatively stable atmosphere inside the command post, he is expected to be the communications star. Virtually everyone on the scene will hear him, so his presence and demeanor should set the tone for the whole show.

REMEMBER: Effective FGC communications techniques—a good FGC image → high worker confidence.

Communication time and space become valuable quantities. Competition can become a major problem. Only one person can talk at a time, but poor timing can prevent everyone from understanding anything. The FGC must regulate who talks and when they talk. He must keep this control and must NEVER allow himself to be knocked off the air.

Communications Guidelines

The following basic techniques will improve fireground communications:

BE SHORT, SPECIFIC, AND CLEAR—Know what you are going to say before you key the mike. Choose precise, short terms, and avoid uncommon or little-used words. Common language and standard firefighting terms are best understood. Operational orders should be specific.

AVOID DISTRACTING MANNERISMS—Use a natural tone; strictly avoid whispering or shouting. Speak in a clear tone, at a normal rate.

PRIORITIZE MESSAGES—Send critical messages first. Maintain radio discipline, avoid informality, and do not interrupt unless you have emergency traffic. (Listen before you transmit).

KEEP MESSAGES TASK-ORIENTED—Indicate a specific assignment or task which outlines what do do, not how to do it. Those receiving the message need to know where to go, to whom to report, what to do, and the desired results.

FOLLOW THE ORDER MODEL—Be certain that the receiver is ready to receive before transmitting the assignment and make sure the message is acknowledged. A brief restatement of the message is far more effective than, "10-4."

SUMMARY

The initiation, maintenance, and control of communications is a basic fireground function of the FGC. Communications serve as the connection between management and personnel.

Fireground communications problems include the lack of SOPs, deficiencies in training, organization problems, equipment problems, and the use of improper communications techniques. Proper prefire training, planning, and readiness can eliminate most of these problems.

The FGC must establish communications to advise all companies of the details of the attack plan to receive needed information. The initial report should explain the conditions seen from the command post. It should include building size, height, occupancy, fire and smoke conditions, the confirmation and designation of command, and the action being taken.

Remember, the standard initial transmission becomes an indication of things to come. Start with an organized beginning. The initial transmission should be short, process information, order specific actions, and provide an accurate status report (progress, completions, and exception reports).

Sector reports should be simple and provide the FGC with positions, progress, and needs. Fire company reports should be more basic and provide the assignment to a task with both location and objective. They should indicate the resources, support, and supervision required. These reports should provide for reassignment once a task is completed.

After an initial assignment is given to a company, there is little need for additional transmissions unless the company is faced with unexpected problems or barriers.

The command post staff must be communications technicians, following a definite communications plan. All command post staff should be prepared to receive, think, and transmit.

In the control of communications, the human factor is the deciding factor. During the course of the operations, the FGC must be heard by everyone on the scene. He must use good communications techniques to project a good command image. He should follow the standard communications guidelines, keeping transmissions short, specific, and clear. He must avoid distracting mannerisms. The transmission must be task-oriented and follow the order model. The messages should be prioritized.

COMMAND DEVELOPMENT

Effective communications is a responsibility of the FGC. Prefire planning and the assurance of properly and continuously trained personnel are key factors in reducing the common communications problems that are associated with the fire scene. Since everyone working the fireground has the potential of hearing FGC communications, the FGC must lead by example. The FGC must know, use, and enforce standard communications guidelines.

The following report card is provided so that you can evaluate your FGC knowledge and skills in classroom exercises, simulations, and on the fire scene.

Fireground Commander Report Card

Subject: Communications

Did the Fireground Commander:

☐ Assure SOPs for communications?
☐ Initiate communications upon arrival?
☐ Communicate conditions as seen from the command post?
☐ Keep sector reports simple, providing Command with positions, progress, and needs?
☐ Keep fire company communications basic, providing the assignment to a task with location and objectives?
☐ Communicate with fire companies in terms of resources, support, and supervision required?
☐ Keep company communications to a minimum after the initial assignment was given?
☐ Utilize the command post staff for communications?
☐ Project a good command image through his communications techniques?
☐ Follow standard communications guidelines during the operation?

3 FUNCTIONS OF COMMAND

Section 4
IDENTIFY STRATEGY, DEVELOP ATTACK PLAN, AND ASSIGN UNITS

MAJOR GOAL

TO USE A SYSTEMATIC METHOD TO MAKE BASIC STRATEGY DECISIONS, AND TO DEVELOP AND INITIATE AN ATTACK PLAN.

OBJECTIVES By the end of this section, you should be able to:

1. State the purpose of the FGC's basic strategy decision. (p. 64)
2. List the major fireground factors defining the offensive/defensive mode. (p. 64)
3. Describe what each of the following should look like:
   A. Offensive mode
   B. Defensive mode
   C. Marginal mode (p. 65)
4. Explain what is meant by the offensive and defensive modes being independent events. (p. 65)
5. List the operational benefits produced by the effective management of the overall strategy. (p. 66)
6. Compare and contrast strategy and the attack plan. (p. 68)
7. List the three tactical priorities and state their importance in the development of an attack plan. (p. 69)
8. State why most fires are more alike than different. (p. 71)
9. List the five basic steps of the attack plan worksheet. (p. 72)
10. State the purpose of the order model. (p. 74)

IDENTIFYING STRATEGY

A major function of the FGC involves translating his evaluation and forecast of conditions into the overall operational strategy. This basic strategy decision serves the critical purpose of determining if the operation will be conducted in the offensive or defensive mode. The development and management of the overall strategy becomes the basis for rescue and firefighting action.

IDENTIFYING THE MODE

The FGC identifies the mode as offensive or defensive through the analysis of an array of fireground factors and their related characteristics. The major factors and questions to consider in identifying the correct mode include:

- **FIRE EXTENT AND LOCATION**—How much and what part of the building is involved?
- **FIRE EFFECT**—What are the structural conditions?
- **SAVABLE OCCUPANTS**—Is there anyone alive to save?
- **SAVABLE PROPERTY**—Is there any property left to save?
- **ENTRY AND TENABILITY**—Can forces get in the building and stay in?
- **VENTILATION PROFILE**—Can roof operations be conducted?
- **RESOURCES**—Are sufficient resources available for the attack?

While it is a major responsibility of the FGC to decide on strategic mode, the entire firefighting team must be able to define, identify, and understand both the offensive and defensive mode. This is an absolute necessity if all the players are to react correctly and efficiently to the FGC's strategic decision. The process must be simple and cannot be a mystery if everyone is going to do an effective job and survive the experience. There is an easy test of understanding. Every firefighter

FIGURE 3.4.1: The operational mode should be easily identifiable to all firefighters.

should be able to look at the operation and identify the basic strategy. If your firefighters cannot identify the strategy, something is wrong.

In determining the strategy, the FGC also establishes the overall goals for the fireground operation. He decides where and when his troops will engage the fire, where they will attempt to stop it, and what the final outcome should be.

Offensive Operation

During an offensive operation, fire conditions will allow an interior attack. In this situation, hand lines are extended into the fire area to support the primary search and to control the fire, while related offensive support activities are provided to clear the way for the attack. This mode is aggressive and quickly moves to "blast" the fire from the inside and extinguish it.

Defensive Operation

During the defensive operation, fire conditions prevent an interior attack, so large exterior fire streams will be placed between the fire and the exposures to prevent fire extension. This mode is a heavy-duty, cut-off oriented approach. It may include operating exterior streams around a large or unaccessible fire area that is essentially burning itself out. During active defensive operations, perimeter control becomes critical since firefighters should not enter the fire area. The FGC concedes property to the fire and decides where the cut off will take place.

Marginal Operation

A difficult and dangerous situation on the fireground occurs when conditions are near the end of the offensive scale and at the beginning of the defensive scale. The FGC may initiate a cautious offensive attack while constantly re-evaluating conditions and the effect the attack is having on the fire. When the FGC first makes a strategic decision, he does so knowing that there may be changes. He has to extend, manage, and control the entire operation within the context of the basic strategy he selects, always prepared for changes.

The offensive and defensive modes are independent events. Effective fire operations (safe, sane, and predictable) are conducted either on the inside or the outside of the building. Any mixture of the two basic modes begins to set the stage for loss of life and property.

The crews fighting the fire must depend on the FGC to control the strategy, knowing they will either be in the offensive or the defensive mode and not a mixture of both. If the inside guys are pulling a two-inch attack line into the building while the outside guys are operating ladder and wagon pipes into the same interior space, effective and safe firefighting has ended. The guys on the inside will soon learn that the outside guys always win and that the FGC is not in control of the fireground. Marginal does not mean both modes in operation simultaneously.

OPERATIONAL BENEFITS

Effective management of the overall strategy by the FGC provides the following benefits:

- Structures decision making and evaluation
- Standardizes understanding and communications
- Supports fireground safety
- Provides operational control
- Improves overall effectiveness.

Structuring Decision Making and Evaluation

The FGC performs a critical function as he manages the strategic process. He starts that function at the beginning of operations by making the initial decision to conduct an offensive or defensive attack. He must then continue to manage the strategic function by keeping the attack plan current in relationship to the inside/outside process. As conditions change, he must be prepared to adjust the strategy and the related plan—particularly if conditions worsen. When active fire conditions become more intense and widespread, and the fire keeps winning, the attack plan must shift from offensive to marginal and eventually to defensive.

The offensive-defensive index provides a standard way to evaluate fire conditions and the effect of firefighting. This index allows for an effective adjustment in the overall operational plan. The system depends on the FGC to decide if firefighters will operate inside or outside the fire building or area. Responsibility is fixed on the FGC to continually re-evaluate that decision. When necessary, he should assemble an effective staff to assist with strategic decisions, develop a hard-hitting plan, and create an adequate fireground organization.

Standardizing Understanding and Communications

Developing and maintaining the overall strategy is not only a major FGC function, it is a definitive agent. This inside/outside decision sets the stage for the entire operation and influences how the basic fire fight will be conducted. The FGC must always be able to define operations within the offensive/defensive framework. The strategy statement is the quickest and most descriptive proclamation of the fire conditions and the actions needed to match those conditions.

This strategy definition serves the purpose of providing a simple and straightforward framework for everyone on the firefighting team to develop a common understanding. "Offensive" means a quick, vigorous interior attack; "defensive" means go for the big guns and watch out for falling walls. "Marginal" means attack carefully, evaluate quickly, and have a fast way out.

The environment on the fireground is fast and dirty. Common understanding facilitates quick communications and effective action and reaction. When the FGC says, "Get out, we're going defensive," the action must be swift. Some FGCs act as if they have the time to send very specific directions to the Interior Sector by way of registered letter. If everyone knows what mode they are in and what it means to be in that particular mode, a simple command can quickly achieve results.

Supporting Fireground Safety

The quick, clear, and concise decision to operate either offensively or defensively serves as the practical and fundamental basis for firefighter safety and survival. This approach to managing the overall strategy requires the FGC to evaluate current conditions and forecast future outcomes. Those factors translate into the position assumed by the attack teams.

Much of fireground safety involves the positions that firefighters assume in relation to what the fire and the fire building are doing. This basic positioning process is a very simple but critical one. The FGC's offensive/defensive decision establishes and maintains control of where the fire forces are located and matches what they are doing to their positions. It attempts to keep the fire and the building from killing the firefighters, while the two distinct modes attempt to keep firefighters from killing each other.

Providing Operational Control

In order to establish an overall strategy that will work for the entire operation, the FGC must approach initial evaluation and attack planning in a rational and systematic manner. This starts the operation in a well-organized fashion and sets the whole stage. Establishing the overall strategy provides a basic and standard structure that assists everyone on the fireground to understand the overall plan and their relative position and function within that plan.

The basic offensive/defensive approach provides the starting point for understanding the planning and control process. The essence of fireground control by the FGC simply involves the ongoing ability to direct where the firefighters are located and what they are doing. In a very practical sense, if the FGC can initially place his firefighters, move them, and change what they are doing, he has the most important level of control. If he is not able to change the location and function of his resources, the operation is out of control.

Improving Overall Effectiveness

Mobilizing the entire operation under the control of a strong strategic plan produces an efficiency of effort that is able to concentrate on a central objective. The objective of each mode is simple and easy to understand:

- **OFFENSIVE TARGET**—do battle inside the fire area to control the fire
- **DEFENSIVE TARGET**—exterior attack to limit loss and stop the spread of the fire.

Management of the overall strategy has more opportunity for determining overall success or failure than does any other function. Simply, offensive operations are different from defensive operations. Nothing else produces effective, standard, and predictable outcomes any better than a fast determination of the correct mode, strong control of the operation, and the maximum effort by everyone to achieve the objectives (nothing produces poor outcomes any quicker than the opposite). Strategic mode confusion unnecessarily beats up more firefighters and burns down more buildings than any other fireground mistake.

Tactical situations in which interior firefighters who are about to complete an offensive attack are bombarded by the outside big guns eventually will produce the need to mobilize lots of donuts in the Canteen Sector. This simple mode confusion will quickly (almost instantly) reverse offensive conditions, destroy the correct direction of offensive support (particularly ventilation), and will drive the inside crews out. One good thing comes from all of this. It gives the inside guys a chance to talk to the outside guys about their parents.

FIGURE 3.4.2: Mode confusion promotes "social interaction" on the fireground.

Mode confusion is ALWAYS a mistake. Sometimes this mistake is ordered by an impatient FGC who can no longer stand to watch the fire push smoke and pass gas. This FGC will order a premature exterior deluge. True, the precontrol period can be a frustrating time for everyone, however, the management of the overall strategy is designed to keep everyone honest—particularly the FGC.

The concept of fireground control becomes an active and very practical reality when it involves frameworking the overall strategy. There isn't any "blue sky" involved in a FGC sitting in his command post looking, listening, and deciding if the attack is going to be from the inside or the outside and then controlling the operation in that mode. He cannot give away the ongoing responsibility for that decision.

THE ATTACK PLAN

TACTICAL PRIORITIES

The FGC must cause his strategy to "come to life" through the development of an attack plan. This plan must be directly related to the strategy but not a substitute for it. The distinction between the strategy and the plan is simple. The strategy describes the general overall approach of the operation and drives the attack plan. The attack plan provides the tactical assignments required to achieve the strategic objective. The order of development is important.

REMEMBER: The strategy comes first and drives the plan. The plan must follow and match the strategy.

FIGURE 3.4.3: Some people never get it right.

The attack plan is based on the three tactical priorities that will establish the major tactical activities required to extend a complete operation. These tactical priorities identify the three separate functions that must be completed in priority order to stabilize the overall fire situation. They are, in order:

1. Rescue
2. Fire control
3. Property conservation.

This list gives the FGC a set of functions (what to do), priorities (when to do them), and benchmarks (how to tell when each function is completed). Tactical priorities provide the FGC with a simple, short list of major categories to act as a practical 1-2-3 guide during the difficult initial stages of fireground attack planning. Complicated plans and guides tend to break down during this critical time.

Tactical Sequence

Tactical priorities MUST be approached in order. The unique realities of the fireground are that the FGC usually gets one shot at certain activities. In most cases, he has only one chance to launch an initial primary search (otherwise, it isn't a primary search). Often, there is only one chance to launch an initial, interior, offensive attack and only one chance to conserve property that is being damaged. The FGC cannot reverse injury, death, or loss after it occurs. He can only interrupt the sequence leading to these events. Therefore, tactical priorities represent intervention plans in an appropriate order to solve fireground problems.

Even though the tactical priorities are interrelated, they are separate and must be dealt with in sequence. The FGC cannot proceed to the next priority before reaching the objective of the current function. He must focus on completing the current objective. This requires command discipline for there are cases where the needs of the current activity may not be obvious—one priority may not be critical or activities may have to be combined to achieve the objective. The framework of the priority list does keep the FGC straight. He must account for each function in order.

Tactical Approaches

Each of the three tactical priorities require somewhat different tactical approaches from both the command and operational standpoint.

During rescue operations, the FGC is attempting to locate and remove threatened occupants. He must be prepared to write off all property to accomplish this objective. His approach is life-safety oriented, strongly influenced by a compressed and sometimes desperate time frame. He may fight the fire in order to complete the primary search, but he stays in the rescue mode until he receives an "all clear" report.

When involved with fire control efforts, the FGC is attempting to find the fire, cut if off, and put it out. This may require an aggressive, crude, heavy-duty operation if he is to stabilize the fire. It may be necessary to beat up the building to accomplish forced entry, ventilation, the opening up of walls, ceilings, and floors, and to operate nozzles. This is a conscious tradeoff of firefighting damage against fire damage. The simple fact is unless he gets on with putting out the fire, he will soon run out of building.

In property conservation activities, the FGC is attempting to identify and protect the value of all that survived the fire and the firefighting. He is now a conserver, where before he may have been a destroyer. Time is less critical, so operations can be more delicate.

Benchmarks

The objectives of each priority are reflected in the following benchmarks of completion:

> **"ALL CLEAR"**—the primary search is completed
> **"UNDER CONTROL"**—the fire is controlled
> **"LOSS STOPPED"**—property conservation is complete.

The benchmarks are achievement signals that tell the FGC when one basic priority function is completed and the operation can go on to the next major activity. The priority activities offer the foundation of the plan of attack. The benchmark establishes practical objectives that are simple, straightforward, and easy for the FGC to concentrate upon.

While the FGC must satisfy the objectives of each function in sequence, he must be prepared to overlap and mix activities to achieve the current benchmark. Notable examples are situations in which it is necessary to accomplish interior tenability through extensive fire control actions before getting on with the primary search or when there is a need to initiate salvage operations while active fire control efforts are still being extended. In situations that require mixed activities, the FGC must realize that he is in the process of achieving the current benchmark even though something quite different may be in progress. A casual observer may be able to identify the offensive vs. the defensive mode, but he may not be able to state what tactical priority is being sought.

ATTACK PLANNING

There is a natural inclination to think of every fire situation as being different. This mental outlook causes the FGC to consider each fire as a totally new experience. This approach is dysfunctional. The FGC is usually confronted with a fairly standard array of factors on the fireground. The major characteristics of these conditions are listed in Section 2 of Chapter 3. Note that this is a fairly short list. Effective FGCs soon discover that fires are more similar than different. If the FGC can develop a standard approach, and then customize the approach to fit each fire, he will begin to develop a plan, approach, or style that can be refined and built upon.

A standard approach develops progressive ability. Each experience adds on to past experiences and gives the FGC the opportunity to improve. He should seldom make the same mistake twice (you rarely learn much from a mistake the second time you make it). The FGC also can profit from the experience and mistakes of others and deposit these lessons into his own "experience bank." Beware of FGCs who must commit every possible fireground screw-up in order to learn the elements of each situation. They burn up a lot of property while receiving their education.

The FGC is generally trying to achieve the same fireground objectives from one fire to the next. Tactical priorities offer a regular set of hooks on which he can hang activities in order to develop this standard approach. With a standard approach, he can begin to manage every fire in the same basic manner (a blessing for those who serve under him!). The tactical priorities serve as the basis for such a standard approach. The FGC lines up rescue, fire control, and property conservation as targets for both himself and his control forces. This standard attack plan approach should become a habit.

If an operation starts in a piecemeal fashion with each company committing themselves to their own plan, the entire operation will be at a continuous disadvantage. Many times these plans involve doing what feels the best (sensual firefighting). Such a fragmented approach can creep into any fire department's operation simply because so many of the fires it fights are obvious offensive situations in manageable-sized buildings. These situations usually can be controlled by a fast, aggressive, unstructured attack.

This routine business can produce an attack habit that becomes a version of the free enterprise system, with companies acting like independent contractors. They get into the habit of attacking in an uncoordinated manner and usually they put out the fire. They are running on luck, and luck makes everyone dumb. The payday for this set of bad habits comes when they are confronted with a large, complex, or unusual situation and attempt to apply their usual (yahoo) attack procedure. They skip the strategic evaluation, never develop an attack plan or assign units to a plan, and soon discover they can't put out or scare out the fire.

Now, some tired, late-arriving FGC must inherit this mess and somehow develop an attack plan that pulls together the operation to deal with the current conditions. This is how you play catch up at its very worst.

When an attack plan is developed by the FGC, it contains a statement of the tactical approach that will be taken and the assignments required to carry out the plan. The attack plan provides a standard process to start, conduct, and conclude operations within the overall plan. This process involves the following basic steps:

1. Evaluate conditions
2. Develop tactical approach
3. Identify tactical needs
4. Identify available resources
5. Make assignments.

Figure 3.4.4 is a worksheet that outlines the items involved in the basic steps and provides a flow chart for completing the attack plan. The five attack-planning steps involve the major factors required to get the operation moving in an effective manner.

The Evaluation Check List

The attack planning worksheet considers the following factors in the evaluation process:

- The basic fireground factor categories provide a framework of the major items that must be evaluated in developing an attack plan.
- Safety factors outline the basic action the FGC must monitor to provide firefighter welfare.
- An evaluation of the size and hazard of the structure(s) and the fire are broken down into standard categories. These categories establish a perspective of how big the operation will be. This basic evaluation determines the resource level that will be required to stabilize the incident.
- The overall strategy establishes the basic inside/outside attack position.
- The 1-10 fire stage scale describes current conditions and is used as a framework to forecast future stages and outcomes.
- Tactical priorities establish the basic tactical targets in the correct order.

Tactical Approach

This statement provides a simple description of the basic tactical approach to be taken:

"Fast interior attack with hand lines to support complete primary search and rescue operations, and control fire in the room of origin. . . ventilation and check for attic extension will be provided along with attack operations. . . salvage operations will follow fire control."

Tactical Needs Translated into Tasks

The FGC must identify the major tactical needs that must be completed. These needs become the basis for assigning specific tasks to companies, such as:

1. Interior hand line attack
2. Complete primary search
3. Control fire
4. Check attic
5. Initiate required salvage.

[1] EVALUATION CHECKLIST

Fireground Factors
- ☐ Fire
- ☐ Building
- ☐ Occupancy
- ☐ Life Hazard
- ☐ Arrangement
- ☐ Resources
- ☐ Special Circumstances
- ☐ Safety

Fireground Safety
- ☐ Full Protective Equip.
- ☐ SCBA
- ☐ Collapse Evaluation
- ☐ Crews Intact
- ☐ Escape Routes
- ☐ Effective Positions
- ☐ Relief Crews
- ☐ Safety Sector

Size & Hazard
- ☐ Light Weight
- ☐ Middle Weight
- ☐ Heavy Weight
- ☐ Super Heavy

Overall Strategy
- ☐ Offensive
- ☐ Marginal
- ☐ Defensive

Fire Stage

1	2	3	4	5
☐	☐	☐	☐	☐

6	7	8	9	10
☐	☐	☐	☐	☐

Tactical Priorities
- ☐ 1—Rescue
- ☐ 2—Fire Control
- ☐ 3—Property Conservation

[2] TACTICAL APPROACH

[3] TACTICAL NEEDS

1 _____
2 _____
3 _____
4 _____
5 _____
6 _____
7 _____

[4] RESOURCES AVAILABLE	ASSIGNMENT	OBJECTIVE
E—		
E—		
E—		
E—		
L—		
L—		
R—		
S—		
—		
—		

FIGURE 3.4.4: Worksheet.

Resources Available

The worksheet requires the FGC to review available resources. He must identify the types of units responding to the incident and list the involved companies in a standard manner. This simple "fill-in-the box" listing gives him a quick reference of the available units. A comparison of the units that are available with the tasks that must be completed provides a basis for determining the number and type of companies that will be required. This matching approach of tasks-to-workers becomes a practical basis for managing resources.

Assignment

The FGC's timing, when making specific assignments to specific companies, is an important factor in effectively integrating each unit into the attack plan. Ideally, the FGC should give assignments to arriving companies as they report to their staged (Level I) positions. Staging procedures are designed to provide a standard system to put companies to work in an ordinary manner. The system should be designed to assist both the FGC and the responding companies in getting the right assignment at the right time.

The primary objective of staging is to prevent companies from rolling directly into the middle of the incident and then becoming totally entangled in confusion and congestion before the FGC can give them an assignment. Using the staging report to trigger the assignment from the FGC eliminates giving the company an order while they are en route to the scene. Without staging, a company can easily become lost in the system.

The FGC must match task assignments to the basic capabilities of each unit. Everyone can perform search and rescue, engines manage and supply water, ladders clear the way and open up, medics provide emergency care, and rescue squads do whatever is needed. The FGC must analyze and assign tasks based on the general profile of needs and capabilities. This mix and match approach mobilizes everyone within the attack plan and takes maximum advantage of the different capabilities of the various types of units.

The FGC can exercise an effective level of command only to the extent that he can develop and deliver clear, understandable orders. These orders are the link between the evaluation/decision process and effective action. The FGC assigns units by giving orders.

A major factor in managing fireground orders is the ongoing relationship between the participants. There must be an effective plan to standardize the actions required of both sender and receiver. There must be standard information and a mutual understanding. The order model provides the basic plan for processing orders on the fireground.

THE ORDER MODEL

The purpose of the Order Model is to provide a set method for processing orders given on the fireground. To be of any use, it must be understood by all those operating at the fire.

The Order Model requires the FGC to begin by identifying tactical needs. Before these needs become orders, the FGC has to translate them into tasks.

THE ORDER MODEL

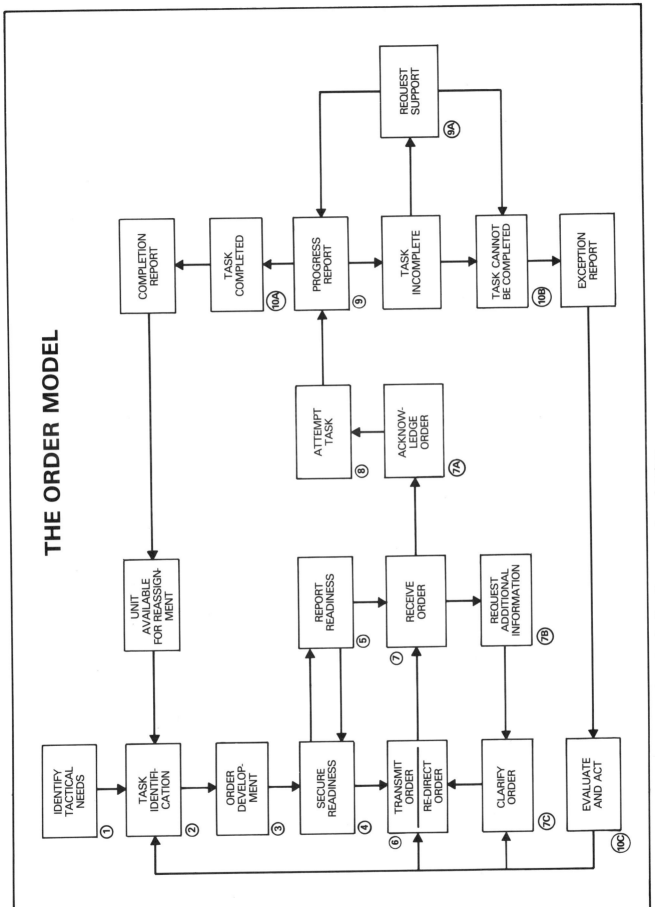

FIGURE 3.4.5: The Order Model is used to standardize the processing of orders.

75

STEP	PERFORMED BY	PROCESS	COMMUNICATION
1. IDENTIFY TACTICAL NEED	FGC	NEEDS OFFENSIVE ATTACK ON 2ND FLOOR	
2. TASK IDENTIFICATION	FGC	ENG 3 IS AVAILABLE. WANTS ENG 3 TO ADVANCE ATTACK LINE TO 2nd FLOOR	
3. ORDER DEVELOPMENT	FGC	TRANSLATE TASK INTO A SPECIFIC ORDER	
4. SECURE UNIT AVAILABILITY	FGC	MAKE SURE ENG 3 IS READY TO RECEIVE	"COMMAND TO ENG 3"
5. REPORT READINESS	ENG 3	CONFIRM READY TO RECEIVE AN ORDER	"ENG 3"
6. TRANSMIT ORDER	FGC	TRANSMIT ORDER	"ENG 3 TAKE A LINE TO THE 2nd FLOOR AND MAKE A DIRECT ATTACK ON THE FIRE"
7. RECEIVE ORDER	ENG 3	7A. ACKNOWLEDGE THE ORDER	"ENG 3 COPIED-TAKING A LINE TO THE 2nd FLOOR AND MAKING A DIRECT ATTACK ON THE FIRE"
		— — — — — OR — — — — —	
	ENG 3	7B. REQUEST CLARIFICATION OF DETAILS OR ADVISE COMMAND THAT ENG 3 CANNOT PERFORM TASK	"ENG 3 TO COMMAND — DO YOU WANT US TO GO IN VIA THE INTERIOR STAIRS OR BY LADDER TO THE BALCONY?"
	FGC	NOTE: FGC MUST CLARIFY THE ORDER, AMEND IT, OR MAY HAVE TO REDIRECT IT TO A DIFFERENT COMPANY.	"ENG 3 GO THROUGH THE INTERIOR"
	ENG 3	CONFIRMS RECEIPT	"ENG 3 COPIED-TAKING AN ATTACK LINE TO THE 2nd FLOOR VIA THE INTERIOR STAIRS"
8. ATTEMPT TASK	ENG 3	EXTENDS EFFORTS TOWARD COMPLETION OF TASK	
9. PROGRESS REPORT	ENG 3	PROVIDES BRIEF PROGRESS REPORT TO FGC:	"ENG 3 TO COMMAND" (FGC "COMMAND TO ENG 3") "ENG 3 IS IN POSITION. WE HAVE WATER ON THE FIRE"
		— — — — — OR — — — — —	
		[FGC MUST REACT TO NEGATIVE REPORT BY ASSIGNING ADDITIONAL RESOURCES OR BY CHANGING THE TACTICAL PLAN.]	"ENG 3 CAN'T MAKE THE 2nd FLOOR. WE NEED VENTILATION AND A BACKUP LINE"
10. TASK REPORT	ENG 3	ADVISE FGC THAT TASK HAS BEEN COMPLETED OR CANNOT BE COMPLETED	"ENG 3 TO COMMAND" (FGC "COMMAND TO ENG 3") "WE HAVE THE FIRE KNOCKED DOWN AND WE ARE READY TO BEGIN OVERHAUL"
		— — — — — OR — — — — —	
		[FGC MUST REACT TO NEGATIVE COMPLETION REPORT — NEW ORDERS MUST BE DEVELOPED]	"THERE IS TOO MUCH FIRE UP HERE. WE ARE BACKING OUT"
11. UNIT AVAILABLE FOR REASSIGNMENT	ENG 3	ADVISE FGC THAT COMPANY IS AVAILABLE FOR NEW ASSIGNMENT	"ENG 3 IS READY FOR REASSIGNMENT"

FIGURE 3.4.5A: Order Model Action Chart.

The knowledge of available resources is critical to the order model. Before giving an order, the FGC secures the available status of the company. Once the company acknowledges its readiness, the FGC can give the order. The company must then acknowledge the receipt of the order. A brief restatement of the ordered task sent back to the FGC is the best acknowledgment.

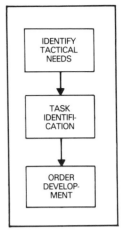

FIGURE 3.4.6: Before an order can be given, it must be developed.

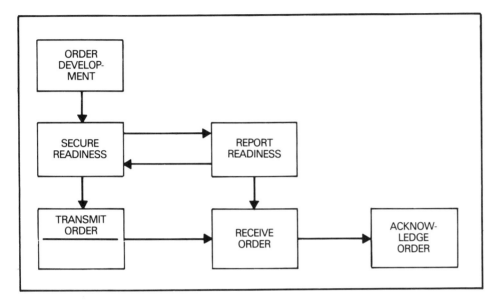

FIGURE 3.4.7: The FGC must know the status of a unit and the order given to that unit.

Orders will not always be clearly understood every time. The receiver must request additional information or clarification if the order is not fully understood, and the FGC will have to translate or redirect the order. He must run his operations so that the company feels comfortable in making this request. Otherwise, a delay may occur or the wrong action may be taken if there is fear of admitting that the order has not been understood.

The FGC must receive acknowledgment of the order after clarification and confirm that the right order was received and understood.

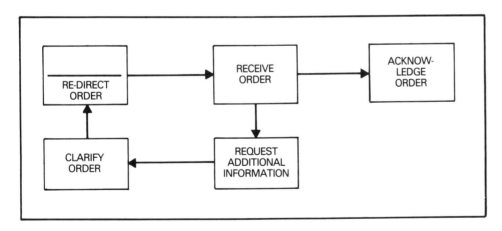

FIGURE 3.4.8: There must be a method to clarify orders.

Once an order is acknowledged, the company attempts the task. They must issue brief, incremental progress reports. During such reports, they should include their requests for support activities, additional personnel, apparatus, and water or other actions needed to accomplish the assigned task.

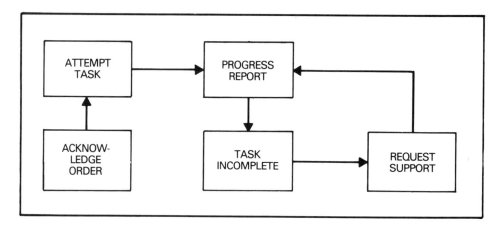

FIGURE 3.4.9: Progress reports must be transmitted to the FGC.

If the ordered task cannot be completed, the company must inform the FGC. This requires an exception report. The FGC will have to evaluate this report and act. This may involve clarification or amendment of the assigned task or the re-evaluation of the attack plan and task identification by the FGC. The company may be reassigned or given the same task to do with support provided by order of the FGC.

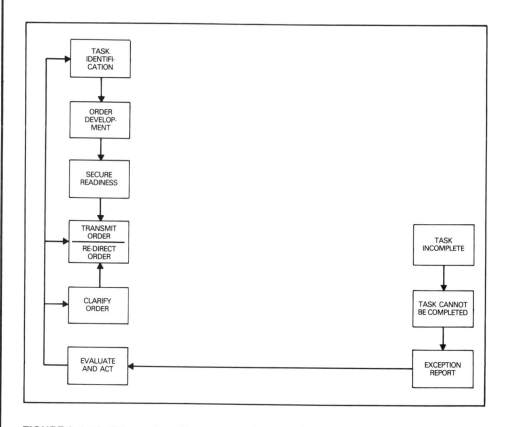

FIGURE 3.4.10: If the ordered task cannot be completed, the FGC must evaluate and act.

When the task is completed, the FGC must be informed. This requires a completion report. The company should let the FGC know when they are available for reassignment.

This system allows the FGC to keep track of the tactical needs that have been accomplished and identify the tasks which are still pending under current conditions.

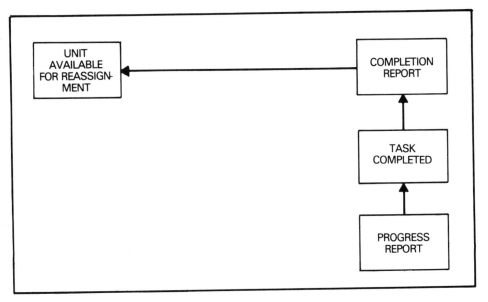

FIGURE 3.4.11: When the task is completed, the company must inform the FGC and report readiness. The FGC must reorder as needed.

SUMMARY

The FGC must be able to translate his evaluation and forecast of conditions into the overall strategy. This decision determines if the attack is to be offensive or defensive and is the basis for all future actions on the fireground.

To define the mode, the FGC considers the fire extent and location, fire effect, savable occupants, savable property, entry and tenability, ventilation profile, and resources. Once he decides on the strategic mode, the FGC must manage within the framework of that decision.

There should be no question as to which mode is being employed. The offensive mode should be a quick, aggressive, interior attack. The defensive mode should be an exterior attack with streams placed between the fire and the exposures.

The offensive and defensive modes are independent events. When the situation is at the end of the offensive scale and the beginning of the defensive scale, the operation moves into the marginal mode. This mode requires strong control of operations by the FGC and Sector Officers.

When the FGC effectively manages the overall strategy, he helps to structure decision making and evaluation, standardize understanding and communication, support fireground safety, provide operational control, and improve overall effectiveness.

The FGC manages the strategic process. The offensive-defensive index provides a standard way to evaluate fire conditions and the effect of the firefighting. This allows for the effective adjustment of the operational plan. Determining the strategy is a standard event. The FGC's strategic definition (offensive/defensive) provides a simple framework for everyone.

Firefighter safety and survival are based on the FGC's clear decision to operate either offensively or defensively. The basic positioning of the firefighers is critical.

When the FGC establishes the overall strategy, he does so in a rational and systematic manner. This starts the operation in a well-organized, controlled manner. The offensive/defensive approach provides the starting point. The strategic plan provides an efficiency of effort, with everyone concentrating on a central objective.

The FGC's strategy comes to life when he develops an attack plan. This plan is based on the tactical priorities, carried out in their correct order. These priorities are rescue, fire control, and property conservation. The benchmarks for each, in order, are "all clear," "under control," and "loss stopped." Each benchmark indicates that a basic priority function has been completed.

Strategy describes the general overall approach of the operation. The attack plan provides the tactical assignments needed to achieve the strategic objective. Strategy drives the attack plan.

The FGC is generally confronted with a standard array of factors on the fireground. Fires are more similar than different. The FGC must learn from his experiences and the experiences of others. He will not do so without attack planning. The attack planning provides a statement of the tactical approach and the assignments. It is a standard place to start, conduct, and conclude operations.

Through the use of a worksheet, the FGC can evaluate conditions, develop action plan statements, identify tactical needs, identify available resources, and make assignments.

Assignments should be made following the Order Model.

COMMAND DEVELOPMENT

It is essential that you learn to follow rational, standard procedures to develop an overall strategy. Your strategy will drive the attack plan. It must describe the general overall approach.

Defining offensive/defensive mode is a critical function of an FGC. To do this quickly and effectively, you must develop your skills in collecting information on and evaluating fire extent and location, fire effect, the primary search, determination of savable property, entry and tenability, ventilation profile, and resources.

Learn to use an attack plan worksheet. Practice with simulations and exercises until you are well-skilled in its use. Not only will it help you to operate in a clear, well-controlled manner, it will improve your ability to learn from your mistakes and the mistakes of others.

The following report card is provided so that you can evaluate your FGC knowledge and skills in classroom exercises, simulations and on the fire scene.

Fireground Commander Report Card

Subject: Identify the Overall Strategy, Develop an Attack Plan, and Assign Units

Did the Fireground Commander:

- ☐ Define the offensive/defensive mode?
- ☐ Manage within the mode?
- ☐ Develop the overall strategy in a rational and systematic manner?
- ☐ Develop a related attack plan?
- ☐ Consider tactical priorities?
- ☐ Utilize an attack plan worksheet?
- ☐ Follow the Order Model?
- ☐ Provide effective communications?

3 FUNCTIONS OF COMMAND

Section 5
DEVELOPMENT OF FIREGROUND ORGANIZATION

MAJOR GOAL

TO DEVELOP AN EFFECTIVE FIREGROUND ORGANIZATION USING THE SECTOR SYSTEM TO DECENTRALIZE GEOGRAPHIC AND FUNCTIONAL RESPONSIBILITY.

OBJECTIVES By the end of this section, you should be able to:

1. State the relationship between the size and structure of the organization and the companies operating on the fireground. (p. 84)
2. Define "sector." (p. 85)
3. Explain how sectoring delegates responsibility. (p. 85)
4. List the three basic operational levels, stating which officers operate each. (pp. 85-86)
5. List five advantages of sectoring. (pp. 86-87)
6. List four situations requiring sector assignment. (p. 87)
7. State the seven responsibilities of the Sector Officer. (p. 90)
8. Give three examples of geographic sectors. (p. 90)
9. Give at least five examples of functional sectors. (p. 92)

FIREGROUND ORGANIZATION

DELEGATION

The fifth basic fireground function is the rapid development of an effective fireground organization to manage operations as the attack plan is implemented. This structure is the link between the command level of the FGC and the action level of the firefighters.

A significant problem occurs when the FGC requests and assigns additional companies at a rate that exceeds the development of his fireground organization. In short order, he will become overloaded with the details of managing a large number of companies scattered over a large area. He will soon find himself in the odd situation of being overwhelmed, yet still in need of more resources to do the job. (This places the FGC in roughly the same position as Custer calling for more Indians.)

FIGURE 3.5.1: When the FGC is overwhelmed, calling for more resources is like Custer calling for more Indians.

Command must constantly match and balance the size and structure of the organization with the number of companies operating on the fireground. This balance is maintained by the assignment of sub-commanders to geographic (area) and functional activities. These people are designated as Sector Officers. They answer directly to the FGC.

FIGURE 3.5.2: There must be a balance between the number of companies and the size of the organization.

The Role Of Sectors

A sector is a smaller, more manageable unit of fireground command. Sectoring is a standard system of dividing fireground command into these smaller units or blocks.

The standard management function of delegation is achieved by the FGC on the fireground through the use of *sectors*. This assignment and the sharing of responsibility and authority become a support mechanism allowing the FGC to divide command, reduce stress, and to maintain control while dealing with increasing operational volume and pace.

Once effective sectors have been established, the FGC can concentrate on overall strategy, attack plan management, and resource allocation. Each of his Sector Officers becomes responsible for the tactical deployment of the resources at his disposal, communicating needs and progress back to Command.

Utilizing Operational Levels

Ideally, a complex incident will include three basic operational levels.

STRATEGY—operated by the FGC, involving the activities necessary for overall operational control, establishing objectives, setting priorities, and allocating resources. This takes advantage of the stationary position of the Command Post.

TACTICAL LEVEL—operated by Sector Officers, assigned to specific areas and tasks by the FGC in order to meet operational objectives. They are responsible for the tactical deployment of resources assigned, evaluation, and communication with the FGC. They supervise directly at the site of the assigned activity.

TASK LEVEL—operated by fire companies, involving the evolution-oriented functions needed to produce task level outcomes. The Company Officers report directly to the FGC or Sector Officer in their assigned area.

Correct use of this three-level organization will increase the collective impact of the entire attack force. The FGC must create the correct arrangement (structure) and the correct amount (scale) for each level if he is to balance management with action.

To develop a balanced fire attack, management must begin at the *task level*—the activities required to stabilize the situation. The FGC must estimate how many companies will be needed, initially and over the entire incident, then build a structure around these educated guesses.

Building an organization from the bottom up places the emphasis on the action level. This is desirable since this is where fire victims get rescued and fires get extinguished. The effective FGC will realize that his function is to support the companies doing these crucial jobs. His command system must match and reinforce the action, not the reverse.

SECTORING

THE ADVANTAGES OF SECTORING

The sector system provides direction and support to the operating units so that they may achieve their objectives safely and efficiently. Utilizing sectors provides the following advantages:

- *Reduces FGC span of control*—divides the fireground into more manageable units.

- *Creates more effective fireground communications*—permits the FGC to exchange information with a limited number of groups and individuals. This reduces overall radio traffic by allowing companies and Sector Officers to communicate face-to-face, freeing the radio for more critical FGC-to-Sector Officer interaction. Communications necessarily become a reflection of the organization; effective communications result from effective organization.

- *Provides a standard system to divide large geographical incidents into effectively sized units*—Sector Officers become evaluation and reporting agents, eliminating the need for the FGC to orbit the entire fire site and allowing him to concentrate on strategy from one position.

- *Provides an array of major support functions*—these are to be selected and assigned according to the particular needs of each situation. The execution and details of these specific operations become the responsibility of the Sector Officers, not Command.
- *Improves firefighter safety*—allows each Sector Officer to maintain control of the position and function of his companies at all times. Sector Officers concentrate on their assigned areas and are in a position to move personnel based on the FGC's decisions.

The use of sectors gives the FGC an ongoing capability to manage any type of incident he may encounter using the same management system, whether it is a fire, major EMS alarm, earthquake, structural collapse, falling meteor, or any combination of the above.

Regular use of sectoring also provides organized "practice" sessions for the players. They have the opportunity to work on small-scale, everyday incidents in preparation for larger, less frequent situations.

ASSIGNING SECTORS

Sectors are assigned based on the following factors:

- *Early fire stage overload*—when the number of assigned and operating companies threaten to overload the FGC's ability to command. Direct control should be delegated to Sector Officers before the FGC's ability to cope is exceeded.
- *Major operation prediction*—when the FGC forecasts that the situation will become a major operation, soon exceeding his span of control. The faster he predicts the build-up, the faster the correct organization can be in place to avoid playing "catch-up."
- *Isolated tactical positions*—when fire companies are operating in tactical positions over which the FGC has little or no control (i.e., complex, interior operations or on a side of the fire he cannot see).
- *Dangerous conditions*—when the FGC must maintain close control over operating companies faced with unusually dangerous conditions. Scenarios involving unsafe structural conditions, hazardous materials, quickly changing conditions, or "one way in-one way out" problems require close supervision and control.

The fireground organization begins with the deployment of the first-arriving units. A regular management system will typically assign companies to key operating and support positions. Usually the first-arriving unit, generally an engine company, will be assigned the most critical tactical position. Later-arriving units will support and reinforce that position.

The combination of this first-arriving engine company, working together with an additional engine company and support company (ladder or rescue squad) can become an *Attack Team* assigned to a specific task or area. The initial-arriving Company Officer will usually become the team leader. When a Command Officer arrives and assumes command, the attack team would generally be given a sector designation. As an operating sector, it becomes an integral part of the fireground organization.

This standard resource build-up provides a regular framework for escalation. When the fire is stabilized, the escalation stops. The whole escalation process is continually moving toward the development of the correct number and types of sectors for the incident.

In order for this system to work there can only be one "playbook" for establishing sectors used by everyone on the fireground. This playbook must be learned and remembered. The FGC and the firefighter should be equally prepared to use whatever sectoring they need and to escalate the system to disaster management proportions when necessary.

Since sectors are assigned based on the needs of a particular situation, sector jobs should be able to be performed by anyone on the fireground, within the limitations of their apparatus, equipment, and special expertise. This includes the assignment of officers as Sector Officers, wherever and whenever needed. The FGC can now develop an effective organization from the beginning with any cast of command and company characters.

Sector Officers

The Sector Officer should be briefed on the overall strategy and tactical objective of his area or functional assignment. Although the number of units assigned to an individual sector will depend upon needs and conditions, three to five companies are a practical span of control for most situations. Part of the FGC's job is to maintain an awareness of the number of companies assigned to each sector and the particular capabilities of that Sector Officer.

The FGC serves as the resource allocator for each incident and, once sectors are established, assigns companies based on resource requests from the Sector Officers. These companies will come from the Level I or Level II staging areas.

Once a company has been assigned to a sector, it is the responsibility of that Sector Officer to get the company into action. In turn, Company Officers will supervise their own crews in performing specific tasks. Company Officers communicate directly with their Sector Officers to describe work progress, resource needs, and other critical factors.

The recommendation to establish sectors should be made by anyone who identifies conditions or positions that require sector operation. Each sector must be commanded by a Sector Officer who may be a Chief, Company Officer, or even a qualified firefighter.

Initial sector command is usually given to a Company Officer who is in a tactical position first. This requires him to act as both a Sector Officer and Company Officer. The FGC should attempt to relieve these officers with later-arriving Command Officers whenever possible, using regular transfer procedures.

Occasionally, a Sector Officer (preferrably a Command Officer) will be assigned to an area sector to quickly evaluate its conditions and needs and then relay them to the FGC. This officer is in a position to supervise incoming companies from the beginning of the operation. Unfortunately, most fires are too "fast and dirty" to permit this ideal operation. Instead, a Sector Officer will usually be assigned to take over operations that have already begun.

LITTLE FIRE

BIG FIRE

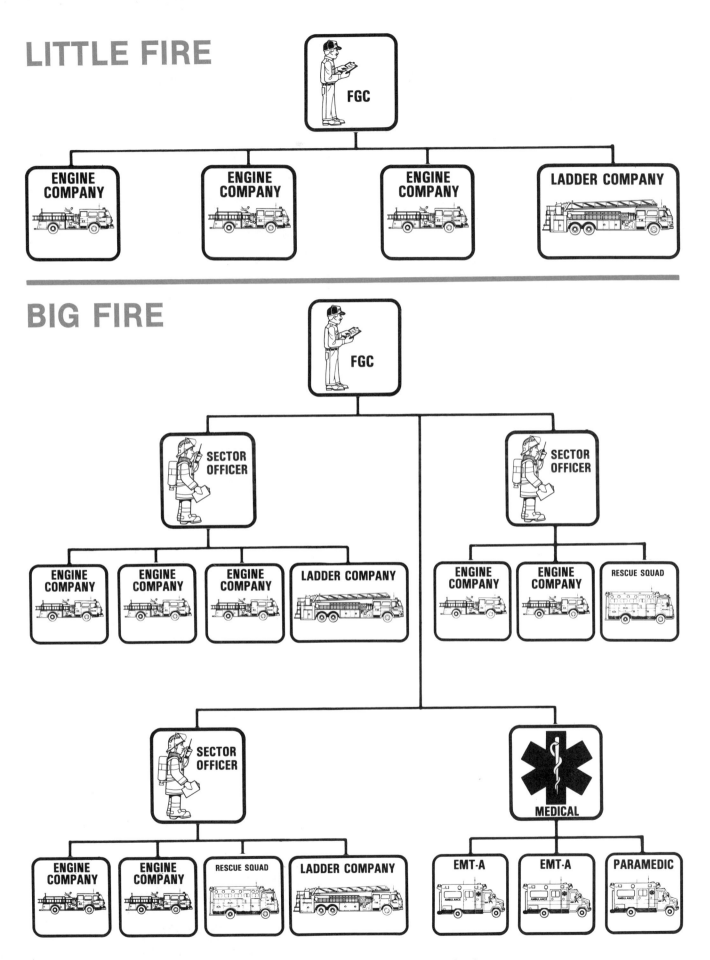

FIGURE 3.5.3: Little fires get a little organization; big fires get a big organization.

FIGURE 3.5.4: Sector officers must be readily identifiable and outfitted with the same gear as their companies.

To efficiently do their jobs, Sector Officers must be outfitted with the same protective gear as their companies and must be in a position to actually supervise the work that is being performed. They should also be easily identifiable, assuming a conspicuous position. When the identity of Sector Officers is not obvious, there is bound to be confusion. Sector vests can be used to reduce this confusion.

Sector Officers are responsible for the following basic functions:

- Directly supervise work in the sector
- Monitor personnel safety and welfare
- Redirect sector activities as required
- Request additional resources from the FGC as required
- Integrate and coordinate actions with other sectors as required
- Advise the FGC of situation status, changing conditions, progress and exception reports
- Decommit companies as operations are completed.

When a company is assigned from Staging to an operating sector, it will be told to what sector and which Sector Officer it is to report. The Sector Officer will be informed by Command as to which particular companies or units have been assigned to his sector. It then becomes the responsibility of the Sector Officer to contact the company by radio or in person to transmit instructions on the specific actions required.

The primary function of the Company Officer working within a sector is to directly supervise the operation of his crew while they perform assigned tasks. As with Sector Officers, they too must advise their supervisor (in this case the Sector Officer) of work progress and, if necessary, additional resources required.

Companies working within a sector should communicate directly with their Sector Officer using non-radio modes whenever possible. The radio will be used for "Sector Officer to Sector Officer" and "Sector Officer to Command" transmissions. Obviously, this plan does not apply to critical, tactical messages or "Emergency Traffic," which can be initiated by anyone at any time.

SECTOR CLASSIFICATION

Sectors should be named using a standard system to clearly identify them for the purposes of operating and organizing. These names are simple one word descriptions of where geographical sectors are or what functional sectors are doing. This sector designation should be used by the Sector Officer in all communications ("Interior Sector to Command").

Geographical Sectors

Geographical sectors are responsible for all general firefighting activities in the assigned area. Some examples include:

- Use of North, South, East, West when the fire area coincides with these directions
- Use of a standard system to number the sides of the fire building and to divide the interior into quadrants designated by the alphabet

- Use of standard landmark nomenclature, e.g., front, rear, roof, and interior
- Use of floor numbers in multistory buildings, e.g., 5th floor, Sector 5.

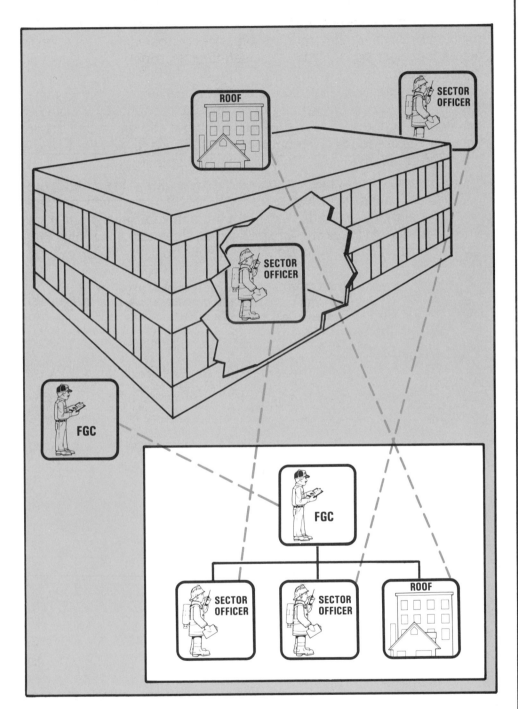

FIGURE 3.5.5: Geographical sectors are responsible for all general firefighting activities in an assigned area.

Geographical sector areas must be well defined to guarantee that no area escapes responsibility. This is a potential problem when companies are operating in large or complex areas. It is more than embarrassing when a fire spreads to an exposure that is not claimed by any sectors.

Functional Sectors

Functional sectors (specialized sectors) are assigned to perform specific tasks or activities which do not necessarily coincide with geographical sectors. The need for these sectors may be easily identified by the geographical sector officers.

On any given incident, the FGC will pick and choose the sector functions he needs using the combination of geographical and functional sectors that best apply. This combination will probably be different for each alarm.

Generally, the FGC will concentrate on the geographic sectors directly involved in firefighting. The majority of functional sectors operate in a support posture on a more routine and automatic basis. Some functional sectors are routinely operated by a single officer, e.g., Public Information Officer (PIO), Safety and Staging.

When a number of active tactical sectors are being directed by the FGC, he may choose to delegate management of these functional sectors to command post staff who will communicate with them on a separate frequency. This ongoing delegation reinforces the reality that the FGC must unload routine details so he can concentrate on critical decisions.

Standard support function areas (sectors) include:

- Safety
- Staging
- Public Information
- Roof
- Resource
- Access Control (Lobby Control)
- Water Supply
- Police Liaison
- Rehabilitation
- Welfare
- Medical
  - Extrication
  - Treatment
  - Transportation
- Hazard

Safety

The Safety Sector is established during active incidents to advise the FGC of existing or potentially unsafe conditions. This sector must be structured to anticipate and deal with unsafe or hazardous conditions. It functions to monitor the condition of personnel, the observance of safety procedures, and the use of protective equipment.

The Safety Sector is responsible for evaluating structural safety, ongoing monitoring of toxic or explosive conditions, and assisting in the management of any special situations that expose firefighters to hazards. It must be highly mobile to monitor the entire fireground, operating with an efficient communications system which works closely with the command structure. Additional personnel may be assigned to provide technical expertise or monitor complex scenarios.

Staging

The Staging Sector assists the FGC with the location, establishment and operation of the Level II Staging Area. A part of this duty is the coordination with police or the Police Liaison Sector to provide for traffic and access control. This sector is also responsible for the inventory and management of regular and specialized equipment in the Staging Area.

The Staging Sector Officer must be visible and accessible. He advises the FGC of equipment and resources available, assigns specific companies to the FGC's requests, and assists these companies in responding to their assignments (routing, direction, access, etc.). This sector may be ordered to communicate directly with Alarm to restock companies or equipment that has been used by the FGC.

Public Information

Media presence on the fireground requires the FGC to establish an effective link through a Public Information Sector. This sector provides a regular place for the media to assemble—away from the Command Post—and a single person to provide the information required for complete and accurate reporting. When the Public Information Sector is regularly used, it builds confidence in both the media and fire department personnel.

The current news business places a major emphasis on television "on-the-spot" reporting as seen on the evening and late news. This production involves camerapersons and reporters who invest in radio scanners and respond to quickly capture all the action. Typically, these men and women are agile and athletic and their work places a premium on showing the excitement. This motivation dictates a strong system to provide all incoming reporters with safe access to shooting and reporting positions.

The best approach is to educate the news media about fire department operations and how they can best fit in, stay out of the way, report well, and survive the experience. A preplanned system to alert the press during the beginning of newsworthy events will further strengthen this relationship. The best relationship is an honest one. Tell them what happened, what action was taken, and the outcome. Needless to say, this process must be done in good taste and with compassion toward the victims and their families.

Roof

The roof of a burning building is a major tactical operating area during most offensive attacks. The Roof Sector is responsible for a significant geographical area (the top of the fire building), but it is also responsible for several specialized functions. The first fire personnel to reach the roof, generally from a ladder company, should establish a Roof Sector and make an initial evaluation of its structural stability. An unsafe roof is an early and critical signal of defensive conditions. The Roof Sector should advise the FGC of the following information:

- Roof structure
- Structural conditions
- Smoke and fire conditions

- Fire wall location and arrangement
- Location of heavy roof objects
- Description of roof actions.

The Roof Sector is responsible for the continuous evaluation of conditions, reporting those conditions to the FGC, directing ventilation, coordinating roof operations with the Interior Sector, and managing the safety of those involved above the fire. Once ventilation is accomplished, the roof should be cleared and, usually, the sector reassigned.

Resource

The function of the Resource Sector is to provide a supply pool for operating sectors and to serve as a "jump-off" point for equipment and manpower close to the action. Once the FGC has determined a need for a Resource Sector, he should designate a location safely outside of the defined fireground, yet convenient enough for support of the other sectors. (During a high-rise fire, the Resource Sector is generally two floors below the fire.) It is the job of the Resource Officer to anticipate equipment and personnel needs, request them through the command system, and keep them ready for action.

This sector may also be used as a personnel staging area close to the fire. In these cases, apparatus will be left in the Staging Area while the companies report to the Resource Sector intact, with full protective clothing and appropriate hand tools. These crews are available for assignment to an operating sector and may be returned to the Resource Sector after the completion of their task.

Access

The Access (Lobby Control) Sector is usually established to control access into a high-rise incident; however, it can be adapted to a variety of tactically complex or unusual situations. Its primary function is to ensure the safety and effectiveness of firefighters by providing a "gate" wherever and whenever direct firefighting traffic control is required.

Access Control Sector personnel directly process authorized firefighters into and out of the operational area and maintain a log of individual entrance and exit times.

Water Supply

The Water Supply Sector represents a concentrated effort directed toward providing a continuous, sufficient water supply. The Sector Officer uses water supply and area maps to first survey the *hydraulic profile* (needed gpm reserve) of the immediate fire area and then to scour for alternate supply sources. Creativity is often the order of the day for Water Supply. The use of portable pumps, large diameter hose, tanker shuttles, and relay pumping operations require reconnaissance to collect all the necessary elements, coordination to set up, and support to operate. Water supply problems are potentially devastating to the FGC—it is important that he plan early for their effective management.

Police Liaison

The majority of large or complex fires require some interaction between police and fire personnel at the Command level. To be most effective, a police supervisor should report to the Command Post and work with the Police Liaison Officer throughout the incident. This sector will coordinate all interaction including:

- Traffic control
- Crowd control
- Evacuation
- Fatalities
- Crime scenes
- Persons interfering with fire department operations.

Once the specific parameters are determined, the police supervisor will assign the details to his subordinates. The control of people, vehicles, and unusual situations, particularly those involving violence, are not in the fire department's job description. (That's why police carry guns.)

Rehabilitation (Rehab)

Long, severe fire operations always affect the physical condition of firefighters. The Rehab Sector provides an organized response to the safety and welfare of all fireground personnel. It should be located in an area remote from the fireground to allow firefighters to safely remove their protective equipment. When faced with severe weather conditions, the Sector Officer should search for a protected area or, when necessary, bring one to the scene.

Ideally, the Rehab Sector should contain a paramedic unit, lighting, an air supply unit, and canteen service. Firefighting crews should be cycled through the sector for medical evaluation, fluids, food, and rest. Crews should be assigned intact and stay together in Rehab until they are ready to be reassigned to new tactical sectors or released from the scene.

All operating sectors should maintain an ongoing awareness of the condition of their personnel and use the Rehab Sector to combat excessive fatigue and exhaustion. It is important that the Rehab Officer maintain a personnel log and coordinate rotation crews with the FGC and other sectors.

Welfare

The function of the Welfare Sector is to monitor the condition of victims and survivors requiring physical and emotional support, determine the resources required to provide for their welfare and then establish liaison with the necessary assisting agencies.

While these citizens may not have been injured, they may have been displaced and require protection from the elements. Other people who may need Welfare Sector services are those who are related to the incident in an emotional way (social support). Since the fire department is not capable of providing such long-term assistance, it must protect these victims until they can be transferred to the appropriate social agency.

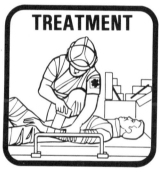

Medical

The fire service is in the most advantageous First Responder position for medical emergencies, particularly those involving many victims. A planned set of Patient Management Sectors allows the FGC to use the overall fireground management system to involve fire companies in an EMS response and integrate those efforts into the existing medical system. These sectors should be routinely established at fires, vehicle accidents, hazardous materials incidents, cave-ins, or any other scenario that produces injuries or medical emergencies. Patient flow is a primary consideration for these tasks.

Medical Sectors may include the following patient management sectors:

- *Extrication*—established in multipatient incidents requiring the extrication of trapped victims. It is responsible for locating, removing, and delivering patients to the treatment area or areas along with the necessary emergency care required for stabilization. The Extrication Sector Officer determines the number, location, and condition of the patients, determines triage requirements, and requests and manages the needed resources. He directly supervises extrication efforts in the area.

  Note—Extrication is an area where rescue and medical overlap. Clear SOPs and a single command system are required to produce a smooth operation and reduce bad feelings as to who is in charge.

- *Treatment*—initiated to provide for complete field triage and patient stabilization. This sector offers continuing care until the patients can be transferred and transported to a medical facility. Its physical location should be away from dangerous conditions, in an area readily accessible for patient entry and ambulance loading. The FGC should quickly move to establish a Treatment Sector whenever the presence of multiple victims is predictable or identified. The function of the Treatment Sector Officer is to supervise the categorization of patients, manage the resources providing treatment, and to coordinate with other sectors and the FGC.

- *Transportation*—the final segment in the patient flow process. It is responsible for the management of ambulances and the allocation of patients to appropriate medical facilities. The Transportation Sector Officer is required to call for the necessary ambulances, set up an ambulance staging area close to the Treatment Sector, determine hospital intake status via Alarm, and coordinate patient transport with "Treatment." Large numbers of patients produce large transportation efforts. Management of this sector requires strong control and adequate resources.

Hazard

The hazard sector is established during incidents where an unusual and/or serious hazard exists. This sector provides the FGC with a Sector Officer who can focus on the nature of the hazard and coordinate the operations required to stabilize the problem. The Hazard Sector Officer is responsible for directing the management of the hazard in a safe and survivable manner. This sector can be established by the FGC if an unusual hazard is present during regular fire or EMS operations or it can be the principle operating sector during a hazmat incident. The hazard sector becomes the regular designation of the hazmat team in departments with such a team. The establishment of this sector signals the entire system that the FGC is dealing with a situation that has a special hazard.

SUMMARY

The fifth basic fireground function of the FGC is the rapid development of an effective fireground organization. From the very beginning, the FGC must match and balance the size and structure of the organization with the number of companies working on the fireground. Delegation is a critical part of making the organization manageable. This is achieved through the establishment of sectors.

The FGC's organization must recognize three operational levels—strategy, tactics, and tasks. Strategy is operated by the FGC, tactics by the Sector Officers, and tasks by the fire companies.

The organization is built from the bottom up, starting with the task level activities that are required to stabilize the situation.

The use of sectors reduces the FGC's span of control, creates more effective fireground communications, provides a system to divide large geographical incidents into smaller areas, provides an array of major support functions, and improves firefighter safety.

Sectors should be assigned in the early fire stages before the FGC's ability to cope is exceeded, when the FGC forecasts a major operation that will exceed his span of control, when operating companies are in isolated tactical positions, or when unusual danger exists. Sectoring provides the organization with a method of escalation that follows the standard build-up from the "first-arriving engine."

Sectors are assigned based on the needs of a particular situation. Sector Officers are assigned and briefed on strategy and overall tactics. Three to five companies are a practical number of units assigned to a given sector. The FGC serves as the resource allocator. Requests come from the Sector Officers.

Sector Officers get the company into action, with Company Officers supervising the crews. These crews perform the specific tasks. Often, a Sector Officer will be assigned to take over operations that have already begun. Anyone acting as both a Sector Officer and a Company Officer should be relieved by later-arriving Command Officers.

The Sector Officer must be easily identifiable, assume a conspicuous position, and be outfitted with the same protective gear as his companies.

The Sector Officer must directly supervise the work in the sector, monitor personnel safety and welfare, redirect activities as required, request additional resources as needed, integrate with other sectors as required, and advise the FGC of situation status, changing conditions, and progress and exception reports.

There are geographical (area) and functional sectors. They should be named using a standard system for efficient identification. Geographical sectors can be named based on compass direction, numerical or alphabetical systems, or standard landmarks nomenclature (e.g., "Roof Sector").

Functional sectors are assigned to perform tasks identified by the geographic sectors. The functional sectors are typically operated in the support posture. Functions that might require sectors include Safety, Staging, Public Information, Roof, Resource, Access Control (Lobby Control), Water Supply, Police Liaison, Rehabilitation, Welfare, Medical, and Hazard.

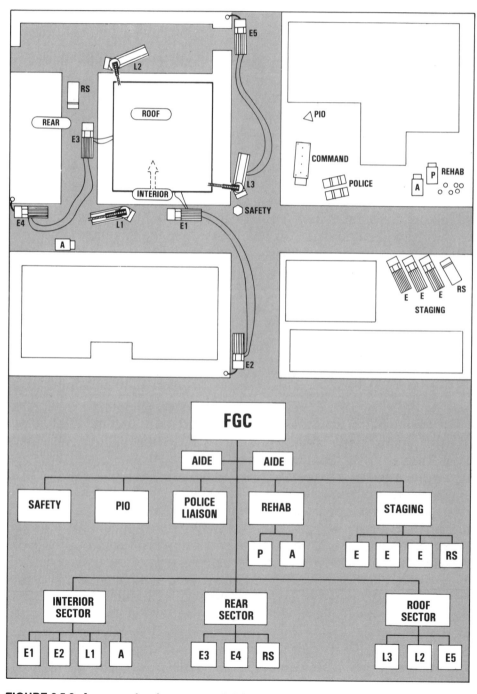

FIGURE 3.5.6: An example of a commercial fire organization.

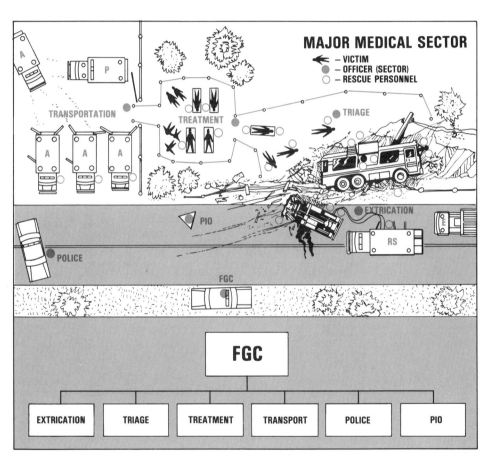

FIGURE 3.5.7: An example of a major medical organization.

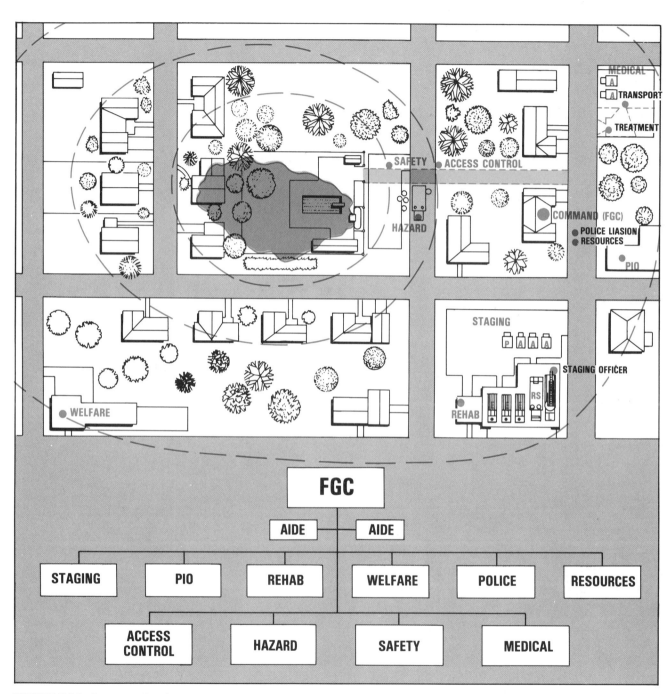

FIGURE 3.5.8: An example of a hazardous materials organization.

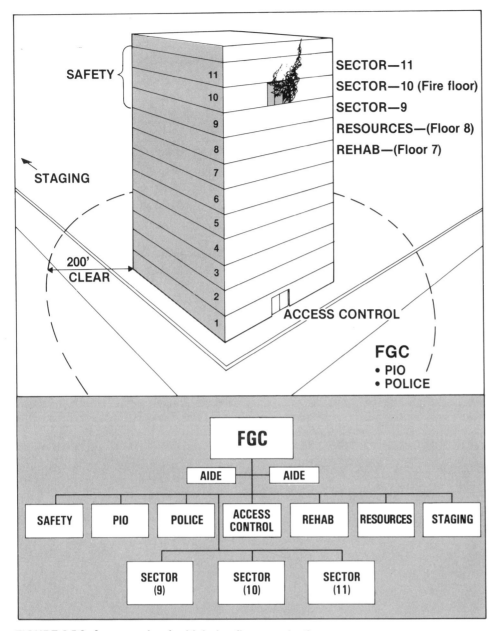

FIGURE 3.5.9: An example of a high-rise fire organization.

COMMAND DEVELOPMENT

To be an effective FGC, you must learn to rapidly develop an organization on the fireground that will support operations as your attack plan is carried out. Quick, efficient, and realistic delegation of authority must become one of your traits. Learn to see the fireground in terms of sectors. Practice doing this for each fire or simulation. Determine your role, your Sector Officers' roles, and your Company Officers' roles as you develop an organization based on sectoring. Make certain that you see each fire in terms of the three basic operational levels (strategic, tactical, task).

Sectors allow the FGC to divide the span of control into more manageable units. This requires you to have a working knowledge of geographical and functional sectors, to be able to forecast when

the span of control will require sectors, and to know your own limitations at an incident.

The following report card is provided so that you can evaluate your FGC knowledge and skills in classroom exercises, simulations, and on the fire scene.

Fireground Commander Report Card

Subject: Developing Fireground Organization

Did the Fireground Commander:

☐ Rapidly develop a fireground organization?
☐ Match and balance the size and structure of the organization with the operating companies?
☐ Forecast the need for sectors?
☐ Accomplish delegation through sectoring?
☐ Utilize geographical and functional sectors as needed?
☐ Correctly name the sectors?
☐ Assign and brief Sector Officers?
☐ Assign units to a sector based on the conditions within that sector?
☐ Assign sectors throughout the operation?
☐ Serve as a resource allocator?
☐ Operate the strategy level?
☐ Evaluate the performances of the Sector Officers?

3 FUNCTIONS OF COMMAND

Section 6
REVIEW, EVALUATION, AND REVISION OF THE ATTACK PLAN

MAJOR GOAL

TO COMPLETE THE STEPS REQUIRED, AND KEEP THE ATTACK PLAN CURRENT.

OBJECTIVES By the end of this section, you should be able to:

1. State one value of arranging command functions in a standard order. (p. 104)
2. List the five command system items that provide a framework for updating fireground activities. (p. 104)
3. List 10 basic evaluation items that are part of an attack plan evaluation. (see Figure 3.6.1)
4. Match the correct FGC action to given attack plan evaluations. (see Figure 3.6.4)
5. List the 10 items on a standard review and evaluation checksheet. (see Figure 3.6.5)

REVIEW, EVALUATION, AND REVISION

The Command System

Standard command functions carried out by the FGC establish an operating framework for tactical action. These functions are arranged in a *standard order* to produce a set of management and operational activities designed to rescue victims and control the fire.

The FGC establishes himself as the overall incident commander, placing himself in an advantageous position for management and leadership. He begins the command process by evaluating conditions, deciding on overall strategy, and developing a matching attack plan.

To initiate this plan, he must take control of the fireground communications process and make assignments to units to go into action. At the same time, it is essential for him to develop a fireground organization that will support the plan and manage the assigned personnel. Once the FGC has reached this point, he must routinely begin to evaluate the effect of his initial decisions.

This approach uses the process of routinely evaluating progress, re-evaluating conditions, fine-tuning the attack plan and, when necessary, making significant changes in the overall strategy and tactics. The FGC must routinely ask himself, "How are we doing—so far?"

A set of *management elements* are used by the FGC at the beginning of the fire to set up operations. These are investments in the regular command system which pay off when the evaluation/revision stage is reached. These elements remain in use throughout the fire, with certain ones being necessary to provide for effective revision.

Updating Fireground Activities

The following are regular command system elements that provide a framework for updating fireground activities during the course of an operation.

CENTRAL COMMAND—The FGC is responsibile for initiating operations, evaluating their effect, and correcting actions when necessary. His entire team must understand every level of operation, incuding the basic functions of the FGC. This common education will prompt everyone to expect an operational evaluation at predictable points for every fire. Evaluation must never seem to be an unusual or unnatural event.

DECENTRALIZED ORGANIZATION—The implementation and operation of geographic and functional sectors provide for the best combination of area command, coordinated action, and feedback from every critical fireground area.

STANDARD OPERATING PROCEDURES—SOPs produce predictable actions and responses by the entire attack team. They also provide the capability to determine, practice, apply, and refine the standard approaches that the organization will take on the fireground. Fire activities that start with standard actions are easier to revise as the operation goes on.

EFFECTIVE FIREGROUND COMMUNICATIONS—The fireground communications process is required to get the operations started, to keep them going, and to provide feedback to match current conditions. Effective communications must be built into every effort and activity. They link together the entire team.

STANDARD ATTACK PLANNING—The elements involved in developing and extending the attack plan provide the structure for evaluating its ongoing effectiveness. The same items included on the attack planning worksheet naturally transfer to the attack evaluation worksheet.

THE ATTACK EVALUATION

The initial fireground effort is based on the FGC's size-up and the standard attack planning elements. Once the basic attack is in place, the evaluation should begin. Figure 3.6.1 lists the categories for an effective evaluation.

This worksheet provides the FGC with a standard set of review items. It serves as a flowchart for completing the evaluation and organizing the required revisions. The FGC must consider every item, asking himself the correct questions to ensure an adequate review and evaluation.

Fireground Safety

Has the FGC provided for the safety and welfare of his firefighters?
Considering fireground safety makes certain that operational effectiveness directly relates to the condition of the workers. No matter how well the operation appears to be going, it is not going well if the fireground is not relatively safe.

This safety factor has to be a concern of everyone on the fireground. The FGC must motivate and lead this concern by:

1. Deciding on the overall strategy
2. Managing the fireground within his offensive/defensive decision
3. Dividing the operation into manageable sectors that concentrate on worker welfare.

In some cases, it may be necessary for the FGC to establish a special Safety Sector to evaluate the tenability of operating positions, to make sure that safety procedures are being followed, and ensure that all firefighters operating within the fireground are always fully protected.

When a violation of safety standards or an unplanned event affects operational safety, it must be the number one priority for review and revision by the FGC. If he is uncomfortable with the location of his firefighters, how they are dressed, or what they are doing, he must correct these things to bring the operation back within a boundary of effective safety. *The FGC absolutely cannot accept a bad safety situation.*

Fire Stage/Strategy

Does the overall offensive/marginal/defensive attack plan strategy match the current stage of the fire?

BASIC EVALUATION ITEM	NEEDED REVISION
1. SAFETY	
2. FIRE STAGE/STRATEGY	
3. 1-2-3 PRIORITY PROGRESS	
4. CORRECT ACTION	
5. LOCATION OF ATTACK	
6. SIZE OF ATTACK	
7. TIMING/AMOUNT OF SUPPORT	
8. ADEQUATE BACK-UP	
9. OPERATIONAL CONTROL	
10. ADEQUATE RESOURCES	

FIGURE 3.6.1: Attack Plan Evaluation Worksheet.

The basic strategy determination provides the basis for effective and safe attack plan development. It also provides the FGC with a practical index to link actions to conditions. Overall strategy management is a major FGC function which must be continually reviewed and adjusted to match current conditions. In other words, if the situation is defensive but firefighters are working inside—pull them out. If it is offensive but firefighters are standing outside—send them in. This is not a complicated concept, but it is extremely important.

1-2-3 Priority Progress

Is the operation moving along the basic order of rescue, fire control, and property conservation?

These activities are the list of essential functions that must be performed as a very simple system. The FGC's plan should reflect the importance of the possible outcomes within each priority and be built around the understanding that it may be necessary to write off property to save lives. Controlling the fire must be a secondary consideration whenever there are people waiting to be rescued. When those rescues are completed, the fire department will do everything possible to limit property loss.

The efficient FGC manages operations around the standard priority benchmarks by directing the right question to the right person, in the right position, at the right time. The essential questions are:

1. *"Is everybody out"* (Has the life safety been secured?). The question seeks an "all clear" report from the Interior (primary search) Sector.
2. *"Are we ahead of the fire?"* (Has the fire been cut off?). This requires a "fire knocked down" or "fire under control" report from the Interior (attack) and Exposure Sectors.
3. *"How does salvage look?"* (Has property loss been controlled?). The FGC seeks a "loss stopped" report from the Salvage Sector.

When the fireground action does not match the correct priority order, the FGC must realign the operation by going back and picking up the missing function. Watching firefighters haul salvage covers into a building that has not been searched is such a signal. Mixing up the order of action or skipping a priority will usually cause major problems later when the uncompleted objective is discovered.

Correct Action

Are firefighting crews applying the correct set of techniques, evaluations, and procedures on the tactical and task level?

Every priority established by the FGC and every basic function carried out on the fireground must have a set of tactical guidelines that promote good fireground practice. Many of these activities involve the mechanical use of tools, equipment, and apparatus. These evolutions make up the collective work done by fire companies and represent the FGC's practical capability to manipulate the fire. Much of this tactical effort involves the movement and application of water along with the extension and coordination of support activities.

Effective firefighting is essentially highly skilled manual labor. The FGC, through the supervision of Sector Officers, must monitor and coor-

BASIC ORDER
1. RESCUE
2. FIRE CONTROL
3. PROPERTY
   CONSERVATION

ESSENTIAL
FUNCTIONS

107

dinate the ongoing application of these manipulative activities to verify that they are indeed working.

Location Of Attack

Are the correct key attack points being covered?

The location of basic attack functions is always a major fireground decision. The fire outcome will usually depend on where the companies are assigned (based on the layout of the fire area), what they are doing, and where the fire is burning. These location decisions are based on:

- Covering the primary search
- Protecting the unburned portions of the fire area
- Cutting the fire off from separate, uninvolved areas.

The FGC reviews and maintains control of the basic operating positions through accurate evaluation, strong organization, and fast communications. When a position is working, the FGC should maintain and reinforce it. When current positions are losing their effectiveness, he must re-evaluate and coordinate position changes.

Size Of Attack

Is the attack large enough to control the fire?

A basic index of effectiveness relates directly to the size of the effort extended vs. the problems faced. This effort involves the ability to extend and protect a fast and complete primary search and also the amount of water that can be applied directly on the fire.

FIGURE 3.6.2: The outcome of a fire depends on relative force.

This entire tactical package is related to the capability of the FGC to deliver firefighters, fire apparatus, and organization in time to interrupt loss (human and property). He is continually involved in a game where the scoreboard shows relative force—if the fire has more force than the FGC, it continues to burn . . . if the FGC can overpower the fire, he gains control of the situation. In every scenario, the FGC must evaluate the need for resources and increase his capability where it is lacking.

Timing And Amount Of Support

Is the fire attack receiving adequate support?

Most fire situations require the removal of barriers that prevent or slow down a direct attack. Assistance is needed for forcible entry, ventilation, and the provision of access. The fire gains a head start directly related to the extent that these barriers separate firefighters and water from the fire. If this separation goes on long enough, the fire simply takes over the fire area, creates a defensive situation, and wins the battle.

FIGURE 3.6.3: Barriers give the fire a head start.

Sometimes this attack effort can produce enough water to float all the ladder trucks away, yet the fire may continue to burn. The problem is to get past all the security and construction barriers and apply water directly onto the base of the fire while it is still in the interior attack, offensive stage. The FGC should take note of the times when there is plenty of water available in charged hose lines but the fire keeps getting bigger. These are the times to ask for feedback from the interior positions and increase and inspire the support efforts to get them into effective action.

Adequate Back-Up

Are resources in place to reinforce the current operations?

Attack plan revision must provide for a tactical reserve in situations that have the potential to expand. These uncommitted firefighters are prepared to support workers in critical positions to provide the additional effort necessary to win. They also may be utilized to cover a new position or function that emerges once the initial attack has begun. The FGC needs to keep a sufficient tactical reserve ready to go to work as soon as the need is identified.

Every tactical position should be evaluated in terms of initial attack and then of back-up. Tactical back-up may include moving up an additional attack line to provide more water to a forward position, providing an additional company to an Interior Sector to strengthen the primary search, or assigning a ladder company to work with an engine company that is blocked by a barrier. Once all the actors are in motion, the FGC should check his tactical worksheet and ask himself the classic question, "What am I going to do if . . .?" When the answer is, "I don't have anyone to do it," he has reached his resource limit.

It is a common fallacy that officers must always assign all available workers on the fireground. This approach overlooks the need for an uncommitted tactical reserve which can be quickly assigned whenever it is needed. The "everybody plays" mentality can congest the work area. It is very easy for the FGC to end up with 17 firefighters working in a 10' by 12' bedroom (usually detected by the sound of SCBA tanks banging into each other). The skill is to balance resources to allow enough workers to do the job, yet keep extra players out of the way until they are needed.

Back-up companies waiting for assignment at a staging position can be plugged into any holes that are detected in the attack plan that routinely occur, beginning with primary tasks and moving to support gaps. These back-up positions become particularly critical when the fire is still burning, and there is still plenty of unburned good real estate left to consume. The clever FGC develops a realistic respect for such potential and automatically lines up a reserve waiting "on the bench" to be sent into the game.

Operational Control

Does the FGC have effective command of the operation?

The following discomforting questions can be used to evaluate whether or not the FGC is actually in command:

- Can he control himself?
- Can he maintain an effective command position?
- Does he know what is going on in all of the critical areas?
- Does he have a plan?
- Will his troops listen and talk to him?
- Will his troops follow his instructions?
- Can he evaluate the effectiveness of operations and make decisions?

- Can he change the assignment, location, and status of his resources?
- Can he call for help to get it?
- Can he enforce his plan before, during, and after the fire?
- Is he responsible for the overall outcome?

"Yes" answers describe the extent to which the FGC is really in control of the fireground. "No" answers describe the extent to which the fireground is actually out of control.

Adequate Resources

Has the FGC balanced resources with tactical problems?

Every review includes an evaluation of the resource level available to the FGC on the scene. Some tactical situations move slowly while others move very fast. The FGC must build a resource profile that intercepts and overcomes the time-temperature production rate of the fire.

Deploying resources ahead of the fire avoids operating in a crisis mode where companies are continually chasing the fire and being driven out by deteriorating conditions. (The objective is to drive out the fire, not vice-versa.)

A critical command function is deciding if the companies assigned to the initial alarm can adequately stabilize the problem. Likewise, this is a crucial factor during the evaluation/revision process. The earlier the FGC makes this evaluation, the more effectively he can manage the resource curve by calling for more companies when they are required. The earlier he makes that decision, the better the chance for these additional units to positively impact the problem.

The quickest way to trigger the response of the closest help, provide for move-ups, and indicate to everyone in the system that the FGC has expanded his response to an incident is to simply request a *standard greater alarm.*

While the FGC can special call any particular unit that is needed, he should resist the temptation to piecemeal assistance by dribbling in one company after another. Usually, the overall resource pool needs to be increased. The extra alarm system can provide a number of companies responding together with a significant collective capacity to expand the response.

Requesting an additional alarm also provides for the response of additional support functions and additional command and company officers who can assist the FGC with the crucial task of building an organizational structure that matches and supports the resource level on the fireground.

When the FGC correctly calls for help early in the operation, he will be placed in the position of having to deal with the new incoming companies at a most difficult time. This is the period when fire conditions are the most active. He is just starting the attack plan and the chaos factor is at its highest.

The unstructured arrival of additional companies, each enthusiastic to get in on the action, can cause havoc with even a well-designed attack plan. The *staging procedures* discussed in Chapter 9 provide a simple system to assist everyone with initial deployment. While these concepts apply in all fire scenarios, they are particularly useful during the fast moving initial period when the FGC is trying to place his companies in the right place to do the right thing.

Staging protocols provide for responding companies to pause in standard uncommitted positions until they receive instructions from the FGC. This gives Command the time to decide who goes where, control communications, and to facilitate the quick integration of all the players in the game plan.

The FGC will be ahead of the game if he identifies the conditions that will typically require additional resources, including:

- The actual or potential life hazard exceeds the rescue capability of the initial assignment
- The actual or potential property protection demands exceed the fire control and property conservation capability of the initial assignment
- All companies are committed and the situation is not under control
- Firefighters are depleted, exhausted, trapped, or missing
- The FGC runs out of or needs more of a special resource
- The commitment of operating companies is not effective; major redeployment is necessary
- The situation becomes so widespread and/or complex that the FGC cannot cope. This requires a larger command organization and sector support
- The FGC instinctively feels uncomfortable with conditions at the resource level (don't disregard fireground hunches).

ATTACK PLAN REVISION

Standard review items allow the FGC to check the progress of the attack plan at various stages and show him where to make any necessary transitions. A transition is a significant adjustment in tactics or strategy which is accomplished smoothly, without a complete disintegration and reconstruction of fireground operations.

When the initial plan does not meet the needs of the situation a tactical and/or strategic transition is required. When the game plan is almost on target, only small adjustments are required. On the other hand, a major transition may require changes in attack size, attack location, or even overall strategy. Smooth, safe, controlled transitions are accomplished by active, coordinated management.

Obviously, it is more effective and much easier on everyone to begin the operation on a realistic basis using a healthy degree of pessimism than it is to start off with a bad plan and stick with it to the bitter end. It is more comfortable to downscale rather than get caught short. A good attack plan allows room for minor adjustments as conditions change. It is always difficult to manage huge transitions that require major orchestration.

Attack plan evaluation must be a routine and continuous process in order to hit a moving target, the fire. The FGC must be able to make transitions as quickly as the need is identified. Some days the transitional process is easy, but sometimes on bad days, even minor tune-ups seem impossible.

Required FGC Action

The outcome options for the FGC range from ''excellent'' to ''get out of town.'' The required command action based on attack plan evaluation can be seen in Figure 3.6.4.

ATTACK PLAN EVALUATION	REQUIRED FGC ACTION
EVERYTHING OKAY	• CONTINUE IN CURRENT DIRECTION • MONITOR ACTION • ADD TO AND REINFORCE CURRENT PLAN AS REQUIRED
MINOR REVISION	• FINE TUNE CURRENT ACTION • CONTINUE TO MONITOR FEEDBACK AS MINOR CHANGES GO ON • PROVIDE SUPPORT AND RESOURCES TO SECTORS AND OPERATING UNITS
MAJOR REVISION	• ADD OR CHANGE SIGNIFICANT ELEMENTS OF BASIC PLAN • ACTIVELY COMMUNICATE (TWO-WAY) WITH SECTORS TO DIRECT CHANGES AND MONITOR PROGRESS • CONSIDER ADDITIONAL RESOURCES TO BACK-UP MAJOR PLAN CHANGES
CHANGE PLAN	• ABANDON CURRENT PLAN — SET UP DIFFERENT APPROACH ON STRATEGY • ESTABLISH NEW ORGANIZATION TO SUPPORT REVISED PLAN • TAKE ACTIVE CONTROL OF ENTIRE OPERATION TO IMPLEMENT OVERALL CHANGE
ABANDON OPERATION	• LEAVE FIREGROUND POSITIONS, REGROUP IN LEVEL II STAGING CONFIGURATION, REFLECT ON THE MEANING OF LIFE, CRITIQUE LAST OPERATION AND LOOK FORWARD TO THE NEXT SATISFIED CUSTOMER
SUBMIT FGC RESIGNATION	• SEEK OCCUPATIONAL COUNSELING FOR MAJOR CAREER CHANGE (*CONSIDER FAST FOOD COUNTER WORK*)

FIGURE 3.6.4: Attack plan evaluation.

ELEMENTS OF REVIEW AND EVALUATION

1. SAFETY

Has the FGC provided for the safety and welfare of the firefighters?

2. FIRE STAGE/STRATEGY

Does attack strategy match the stage of the fire?

3. 1-2-3 PRIORITY PROGRESS

Does operation move from rescue to control to conservation?

4. CORRECT ACTION

Are firefighters applying correct techniques, evaluations and procedures.

5. LOCATION OF ATTACK

Are the correct key attack points being covered?

6. SIZE OF ATTACK

Is the attack large enough to control the fire?

7. TIMING AND AMOUNT OF SUPPORT

Is there adequate support?

8. ADEQUATE BACK-UP

Can current operations be reinforced?

9. OPERATIONAL CONTROL

Does the FGC have effective command of the operation?

10. ADEQUATE RESOURCES

Has the FGC balanced resources with tactical problems?

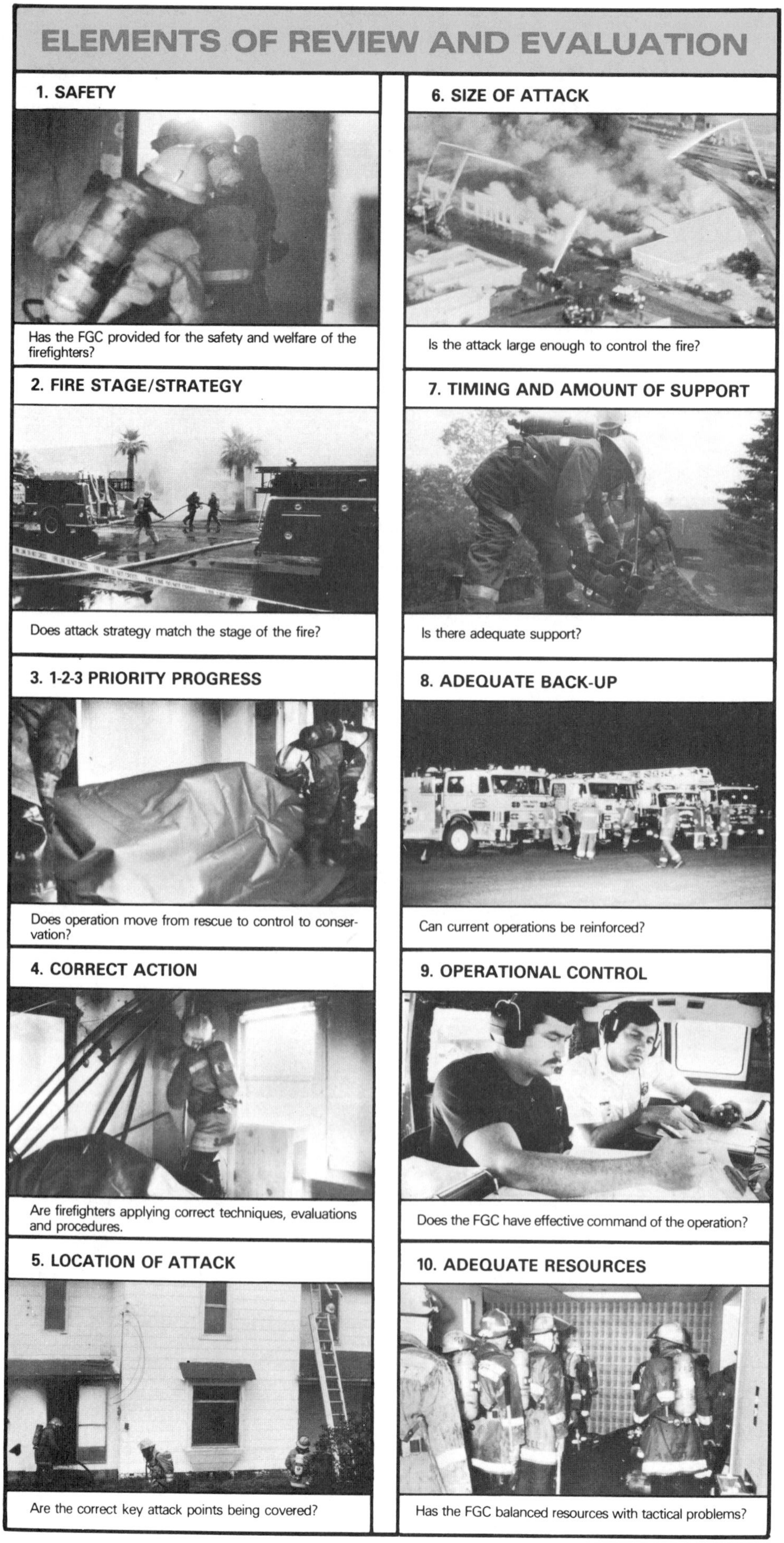

FIGURE 3.6.5: The elements of review and evaluation.

SUMMARY

Effective tactical action requires the FGC to estabish an operating framework based upon a regular series of managerial and operational events. Review, evaluation, and revision must be a natural part of this framework.

Updating fireground activities during the course of action requires that five command system items be present—central command, decentralized organization, standard operating procedures, effective communications, and standard attack planning.

Ten items are a part of the standard review and evaluation process—safety, fire stage/strategy, 1-2-3 priority progress, correct action, location of attack, size of attack, timing and amount of support, adequate back-up, operational control, and adequate resources.

The FGC must provide for the safety and welfare of the firefighters. This safety factor must be a concern of everyone operating on the fireground.

The attack plan strategy must match the current stage of the fire. The operation must move from rescue to fire control to property conservation in the proper order.

Through the supervision of Sector Officers, the FGC must monitor and coordinate activities to ensure that firefighting crews are applying the correct techniques, evaluations, and procedures on the tactical and task levels.

The FGC must make certain that the correct key attack points are being covered and the size of the attack is large enough to control the fire.

The FGC must make certain that the fire attack is receiving adequate support and that there is adequate back-up of personnel, apparatus, and command.

The FGC must have effective command of the operation. He must balance resources with tactical problems.

COMMAND DEVELOPMENT

To be an efficient Fireground Commander, you must be prepared to make review, evaluation, and revision a natural part of every fireground operation. You must learn to carry out the functions of command in a standard order of managerial and operational events. It is necessary to practice by setting up a standard set of managerial elements at the beginning of every fire and using your department's SOPs. Remember, a fire operation that starts with a standard set of actions places you and your personnel in a strong position to revise activities. You must be able to utilize standard attack planning elements and standard review items to effectively evaluate and revise the attack.

The following report card is provided so that you can evaluate your FGC knowledge and skills in classroom exercises, simulations and on the fire scene.

Fireground Commander Report Card

Subject: Review, Evaluation, and Revision

Did the Fireground Commander:

- ☐ Follow regular command system elements?
- ☐ Carry out command functions in a standard order?
- ☐ Review and evaluate each item on the checksheet?
- ☐ Take the required FGC action based on attack plan evaluation?

3 FUNCTIONS OF COMMAND

Section 7
CONTINUING, TRANSFERRING, AND TERMINATING COMMAND

MAJOR GOAL

TO DEVELOP A STANDARD APPROACH TO COMMAND TRANSFER, AND TO OPERATING THE MID-POINT AND FINAL STAGES OF COMMAND.

OBJECTIVES By the end of this section, you should be able to:

1. Relate continuing command to the tactical priorities. (p. 118)
2. List three factors a FGC must consider when deciding how to provide for continuing command. (p. 118)
3. State how the size and complexity of a fire effect continuation of command. (p. 118)
4. Justify a standard fireground management system in terms of continuing command. (p. 118)
5. List four ways in which standard command functions help provide for continuation of command. (pp. 119-120)
6. Tell why a standard system for transfer of command is needed. (p. 121)
7. State why the number of command transfers at an incident must be limited. (p. 122)
8. List three elements of a briefing when a FGC is relieved of command. (p. 123)
9. State the basic transfer rule. (p. 124)
10. Explain why command is usually transferred to an officer of higher rank. (p. 124)
11. List three ways that a FGC can return companies to available status. (pp. 124-125)

CONTINUING COMMAND

LENGTH OF COMMAND

The early stages of fireground operations involve the initiation and implementation of the *basic command system.* This gets the FGC into business and his companies into action.

Once the *basic command functions* are established, the FGC must shift his efforts toward the provision of continuing command until the three basic priorities of rescue, fire control, and property conservation are concluded. This provides regular tactical targets for everyone on the fireground. The time it takes to reach each completion benchmark will add up to the overall time committed to the incident.

The ability of the FGC to effectively conduct his command operation over this time period determines the overall effectiveness of the entire operation. Every tactical situation involves a different combination of the standard tactical elements which effect the length and intensity of operations. The evaluation and planning that go into the attack plan must include a size-up of these elements. If this planning process is not done, the FGC has not really provided for continuing command.

Continuing Command Factors

The FGC considers the following when deciding how to provide continuing command:

LIFE SAFETY CHARACTERISTICS—number, location, and condition of victims

FIRE AREA PROFILE—size, nature, and arrangement of fire load

FIRE CONDITIONS—location, intensity, and the direction and avenue of travel.

Obviously, the size and complexity of a tactical situation will regulate the duration of the operation. The FGC initiates command and ends command. In between, he is continuing command. This in-between period varies the most in regard to size and complexity of the operation. The longer the operation, the more the FGC will depend on and use the basic command functions. More complex operations require a larger command organization and a larger commitment. During this important span of fire fight, it is the FGC's job to continue the essential command functions until the standard rescue and property protection outcomes are achieved.

The Fireground Commander must overmatch fire conditions with the response he can assemble and manage. When operations are short and very active, he must call for adequate resources fast, hit the fire hard, and overpower it quickly. In these cases, the middle period is short yet tough.

When faced with larger operations, the FGC must conduct campaign operations over a longer period of time. He will be effective only to the extent his organization can get ahead of fire conditions and eventually overpower them. The combination of size, high operational intensity, and long duration will test the entire rescue and fire control system in the most severe way.

If the FGC is going to win, he must be prepared to establish and sustain effective operations longer than the fire is able to go on. If he runs out of gas before the fire stops, he loses. The implementation and operation of the command functions covered earlier provide the basic framework for the FGC to extend the direction, control, and support necessary during the entire operation. It would be foolish to develop a command structure that did not last for as long as it is needed.

The command process begins as each function is initiated. The ongoing execution of those activities represents how the FGC "works the system." Eventually, these separate functions combine into a command process that extends and sustains ongoing, overall incident command.

COMMAND FUNCTIONS

The following standard command functions are critical in providing continuing command:

- Command positioning
- Fireground communications
- Strategy and attack planning
- Fireground organization.

FIGURE 3.7.1: It is difficult to extend continuing command from a fire engine cab the size of a phone booth.

Command Positioning

The FGC must maintain a stationary vantage point where he can assemble a small staff of command helpers and increase his communications effectiveness, particularly during campaign-style operations.

Usually, fireground command starts in the front seat of an engine and may move through a series of different command vehicles as they arrive on the scene. The FGC plays with a big advantage when he can escalate and upgrade his command environment to match the physical management needs of the situation.

Fireground Communications

The FGC is able to sustain a continuous command and control capability only to the extent to which he can maintain an effective communications link to his outside world (Alarm) and to his work agents (sectors and companies). This connection provides him with the capability to request additional resources, exchange information, change assignments and locations, and to assemble the information necessary to match actions to current conditions.

Once the FGC gets his office set up, he does most of his business over the radio; therefore, his ability to control his frequencies becomes critical. He must react to anything or anybody who knocks him off the air. The communications process can be his best friend or his worst enemy. Most long tactical festivals become a communications Super Bowl. The FGC will always be challenged to remain on the air.

Strategy And Attack Planning

The basic offensive-defensive strategy decision that the FGC develops to start operations, and the attack plan that emerges from the strategic decision, provide the foundation for continuing command.

The FGC's major job during the middle period of operations is to keep the attack plan working, making the necessary adjustments to continually match the plan with the current conditions of the fire and the way the fire is reacting to firefighting efforts. The *attack planning process* provides a regular system to translate a set of tactical items (fireground factors, size and hazard, fire stage, and tactical priorities) into a sane plan that should always evolve from the simple offensive/marginal/defensive firefighting philosophy.

This attack plan development and use keeps operations going long enough to reach the tactical conclusions included in the plan. The plan provides him with a list of "reasons" for doing the tactical operations necessary during the hard work phases of the fire and a timetable to help decide how long it will take to move operations through this continuing command period. The FGC cannot terminate operations until either all the attack plan boxes have been checked off or the fire wins and burns up those boxes.

Fireground Organization

Effective fireground organization is the major management tool the FGC uses to sustain a safe, coordinated, and standard effort during ongoing operations. The command organization is designed and operated to match the committed resources.

Delegating geographic and functional responsibility to command partners (Sector Officers) reduces the FGC's span of control and increases his ability to operate at a strategic level. This system of a *standard decentralized command* provides off-site managers (away from the FGC) who directly supervise the assigned activities.

The FGC builds his organization by setting up sectors as the fireground operation grows in size and complexity. The arrival of additional Command Officers provides him with "natural players" for these assignments. This standard organization is of particular value during long operations.

Ideally, the FGC would like to arrive at the game in a bus filled with lots of good players, plenty of reserves, and a coaching staff to help him consistently execute plays throughout the game. If the organization has not been developed to adequately support the game plan during all four quarters, plus an occasional overtime, the game may be lost because the fire will never run out of its natural capability, energy, and inclination to injure and kill all the victims and to conduct, and radiate destruction on all the real estate.

FIGURE 3.7.2: If the organization doesn't support the game plan, the game will be lost.

COMMAND TRANSFER

A crisis can exist any time that many potential FGCs arrive on the fireground. The quickest way to destroy effective maneuvers is to allow competing command maniacs to wander through the fire area, each trying to convince the workers that he is the real FGC by shouting conflicting orders. Some orbit clockwise while others move counterclockwise. ALL PRODUCE CHAOS. The end result is a big burn unless the troops can hide from the roving generals long enough to put out the fire. The resolution is a system which allows only one

Mop
602

FGC at a time and defines command transfer rules to be used when passing the baton from the current Commander to the new one.

To provide continuous command, the first fire department unit or officer arriving at the scene should assume command until relieved by a ranking officer, or until command is terminated. The assumption of command is mandatory, although the first-arriving Company Officer may choose to pass command to another officer who will arrive soon after. This process allows a FGC to begin the management functions by providing a strong, direct, and early command. As more companies and command officers arrive, the system is strengthened by matching the level of command to actual needs.

Every fire organization must outline the chain of command within which command will be transferred. This system may be based on a variety of factors that are generally a reflection of local history, culture, and the leadership style of that particular department.

Regardless of the chain of command used, it must be comfortable for the user and work for the organization. The details are not as important as the fact that a standard system is adopted, used, evaluated, and refined based on experience.

The way to designate FGCs during the command transfer process might include:

- The firefighter who got there first
- The highest ranking officer
- The oldest (seniority)
- The most capable
- The person with some special knowledge
- The person with command duty (designated on a rotating basis)
- The toughest (be careful of weapons)
- Social security numbers (ascending or descending)
- The person who lost (or won) the toss.

The system should provide for some practical limit to the number of times command is actually transferred. In most cases, two turnovers will get the level transferred up to the Fire God ranks. More than two transfers often tend to create more confusion than actual command improvement and begins to look like musical chiefs to the observers.

FIGURE 3.7.3: The system for transfer has to be practical and must limit the number of times command is changed.

The following transfer procedure might apply to a medium-sized department with on-duty command staff.

- First-arriving Company Officer automatically assumes command and initiates FGC functions.

- First-arriving Battalion Chief automatically assumes command after completing command transfer procedures and continues FGC functions.
- Division Chief (shift commander) automatically assumes command after completing command transfer procedures during complex tactical situations which have not been declared under control and continues FGC functions. Assumption of command in other situations is discretionary.
- Assumption of command by higher ranking officers (above Division Chief) is discretionary.

The actual command transfer is regulated by a very simple, straight-forward procedure that includes:

1. The current FGC assumes an effective command position.
2. The arriving ranking officer contacts the FGC directly. Face-to-face is always preferable, however, command transfer by radio can be accomplished during fairly simple incidents when the responding officer has copied all command activity made before his arrival. Standard communications are important here.
3. The FGC being relieved will provide a briefing that includes:
   - Situation status—"What have you got?"
   - Deployment and assignment—"What have you done?"
   - Tactical needs—"What do you need?"
   This briefing concludes with a confirmation of command transfer ("I've got it").
4. The use of tactical worksheets to outline the location and status of resources which will assist the transfer process.
5. The regular use of the radio designation "Command" by the current FGC simplifies the entire process. Most command transfers will go unnoticed by the workers. Usually, they don't care who is in command, as long as someone is.
6. Command officers should eliminate all unnecessary radio traffic while responding. Air time during the response period is generally critical. It can be very confusing if every officer with a radio within five miles wants to know what color the smoke is. Basically, everyone should shut up unless they know something that will help the effort.

Transfer And Rank

The arrival of a ranking officer on the fireground does not, in and of itself, mean command has been transferred to him. Command is only transferred when the procedure is completed. The system is designed to keep officers honest. If they outrank the current FGC, they can transfer command from him or they can work under his command. It is also designed to encourage the additional responding command officers to report to the command post area (Level II staging for Chiefs). These officers are available to the FGC for command relief, as Sector Officers, staff assistants, or in any other capacity that assists him. Since engines respond to help engines and ladders respond to help ladders, it only seems fair that chiefs respond to help chiefs.

The response and arrival of ranking officers on the fireground is designed to strengthen the overall effort of the entire team. The pur-

Purpose of Transferring Command is to Improve The Quality of Management + Leadership That in Place

pose of transferring command is to improve the quality of management and leadership that is in place to support the efforts of the workers.

> **REMEMBER: A good basic command transfer rule is: If you can't improve the quality of command, don't transfer it.**

The entire rank structure is built upon the principle of ascending experience and capability. If functioning correctly, it should be the higher the rank, the higher the capability. The longer the fireground management system is in place and the more it is used, the wider understanding of the system spreads throughout the organization. The system can only be effective when everyone on the fireground is familiar with the details that describe standard operations.

It is the responsibility of ranking officers to continually assume a leadership role in removing the mystery from the command system. Managers of the system must establish a standard operation, prepare all the players to operate in that system, provide opportunities to apply it, and extend feedback that reinforces good performances and educates mistakes. *Beware of the officers who say it takes 58 years to learn how to take command—you may not live that long.*

This approach toward senior officer support creates an atmosphere where ranking officers understand they are always responsible for the ultimate outcome but are comfortable in letting a junior officer who is effectively commanding a fire scene continue his command role. In fact, the necessary location, assignment, and status information is usually difficult to exchange during the compressed time period when it is typically done.

In these cases, the ranking officer may choose to move the junior FGC into a more effective command vehicle, join him in that new location, and assist him in completing the command check list. Such situations provide a bright spot for both officers. The young, aspiring chief completes a positive command job that reinforces all the lessons, with his boss sitting next to him, and the senior officer demonstrates his confidence in a junior officer, promoting command growth and personal development. Junior FGC scenarios are stature-building for everyone involved. Command will obviously be considered to be transferred within this context by virtue of the ranking officer being involved in the command process.

TERMINATING COMMAND

When the FGC comes to the end of the tactical priorities, he must decommit the operating companies and terminate command. This phase may range from very simple during light weight and middle weight situations to a long and complicated operation requiring a massive amount of final coordination at the end of campaign operations.

The normal sector system used to get companies into action can usually be applied to place them back in service when their tasks are completed. In other cases, the FGC may have to assign one officer to determine which companies have their crews and gear intact and are ready to leave. He must coordinate this departure with the FGC and Alarm until all units are back in service. The more effectively the process is completed, the quicker companies become available and the shorter the next response time will be.

It makes sense to return companies to quarters based on their fatigue factor. Simply, the companies that arrived first should leave first. This philosophy supplies the workers in the best physical condition to participate in manual labor thereby maximizing the effectiveness of the available human resources. Obvious extensions of this resource management style include Rehabilitation Sectors, automatic rotation of fatigued companies, canteen services, Safety Sectors, and front-end loaders, instead of scoop shovels.

Some organizations do just the opposite and the fire becomes the personal possession of the first-arrivers. It is referred to as "their fire," which defies any sort of basic and sensible personnel conservation. These systems are generally operated by officers who think of long, manual overhaul operations involving large piles of fire-damaged material as character-building experiences. (These same officers tend to act strange during full moon periods.)

As the operation winds down, the FGC can reduce the size of his command structure. He can often reverse the command transfer process and essentially de-escalate command to lower ranking officers. They can be assigned to the companies who will be leaving the scene last or will be maintaining a "watch line" to secure the scene and monitor extinguishment.

SUMMARY

The FGC takes and establishes command. Once the basic command functions are established, he must provide for the continuation of command.

Adequate provisions for the continuation of command must consider how long it will take to reach each completion benchmark of the three tactical priorities.

The FGC must consider certain factors when deciding how to continue command, including life safety characteristics, fire area profile, and fire conditions.

Continuing command is the "middle period" on the fireground. Size and complexity of the fire regulate the length of this time. When needed, the FGC must have an organization that can outlast the fire.

The function of continuing command is strongly related to the tactical priorities. If the FGC is to provide for continuing command, he must start out with a sequential, systematic, standard fireground management system.

Essential to the continuation of command are command positioning (a stationary vantage point), fireground communications (continuous contact with Alarm, sectors, and companies), attack planning and strategy (keep the attack plan going), fireground organization (the organization must be able to support the attack plan).

Failure to have a standard system for command transfer can lead to multiple commanders. The first-arriving department unit or officer should take command until relieved by a ranking officer or until command is terminated. The chain of command for transfer must be outlined as part of every department's organization. There must be a standard system that is comfortable and works. There must be a practical limit to the number of times command is transferred.

When command is transferred, the FGC being relieved should brief the new FGC with a situation status, deployment, and assignment report, and a statement of tactical needs. The new FGC must confirm command transfer.

The arrival of a ranking officer does not mean that command is automatically transferred to him. The system for transfer must be followed. Transfers of command should improve the quality of command. As rank increases, so should experience and ability. Senior officer support of junior officer FGCs is essential to departmental development.

Terminating command comes with the end of the tactical priorities. The sector system used to put companies into action can be applied to place them back in service. An officer may have to be assigned to determine what companies have both crew and gear intact and are ready to leave. The fatigue factor (first to work, first to leave) is useful in deciding the order in which companies leave the fireground.

As the operation draws to an end, the FGC should reduce the size of his command structure. He can delegate certain termination duties to lower ranking officers assigned to companies that will be last to leave the fireground.

COMMAND DEVELOPMENT

Continuing, transferring, and terminating command are functions of the FGC. To carry out these functions properly, you must:

- Know your department's SOPs
- Learn how to use size-up and fireground factors to determine the size and complexity of a fire
- Build an organization that can outlast a fire
- Use tactical priorities when developing an organization that will support the attack plan
- Effectively use communications
- Be able to assume command even after the beginning of an incident
- Gracefully accept being relieved of command
- Learn to benefit from senior officer reinforcement
- Know when and how to reduce the size of your command
- Use practical criteria when you decommit operating companies.

Fireground Commander Report Card

Subject: Continuing, Transferring, and Terminating Command

Did the Fireground Commander:

- ☐ Estimate the length of his command based on the size and complexity of the fire and his available resources?
- ☐ Consider the time for completing each tactical priority?
- ☐ Consider each of the following:
    - ☐ Line safety characteristics?
    - ☐ Fire area profile?
    - ☐ Fire conditions?
- ☐ Develop an organization that stayed ahead of the fire and could outlast the fire?
- ☐ Assume effective command positioning?
- ☐ Develop effective fireground communications?
- ☐ Effectively command his attack plan (did he keep the attack plan going)?
- ☐ Use a standard system or transfer to:
    - ☐ Take over command?
    - ☐ Give up command?
- ☐ Reduce his command organization as the incident came to an end?
- ☐ Efficiently place companies back in service?

4
RESCUE

MAJOR GOAL

TO SAFELY LOCATE, PROTECT, AND REMOVE FIRE VICTIMS.

OBJECTIVES By the end of this chapter, you should be able to:

1. Cite the major goal of rescue operations. (p. 130)
2. State the principle factor influencing the effect of the fire on the victims. (p. 130)
3. List the three categories of fire victims. (p. 130)
4. Define "primary search." (p. 131)
5. Define "secondary search." (p. 131)
6. State the critical factor in the primary search. (p. 131)
7. State when the FGC should routinely initiate a primary search. (p. 131)
8. State the term used to indicate the completion of the primary search. (p. 132)
9. List the four critical factors of a basic rescue size-up. (p. 133)
10. List the three fire stages and describe the basic rescue approach to each. (p. 134)
11. List the order of victim rescue priorities. (p. 137)

PRINCIPLES OF RESCUE

BASIC CONSIDERATIONS

Rescue operations are the most difficult and potentially confusing of all fireground activities. They require fast, effective decisions from the Fireground Commander and strong, aggressive action from the control forces. The FGC must continually recognize the need for adequate, effective rescue operations and the fact that life safety is the number one reason for initial action and additional alarms. As with all fireground operations, SOPs play a critical role in rescue activities.

Generally, a fire will begin to affect its victims before any fire forces arrive. Initially arriving units will often find people already in need of some type of rescue. This is a time for quick, accurate analyses and decisions.

Classifying Fire Victims

There are several factors which influence the effect of the fire on the victims; however, the principle factor is the location of the victim in relation to the fire. In other words, the closer the fire is to the victim, the greater the danger. The fireground realities of victim-to-fire position determine how the FGC plans and extends the rescue operation.

The arriving FGC can classify fire victims according to one of three categories:

1. VICTIMS ALREADY OUTSIDE THE FIRE BUILDING. Usually, conscious and able occupants have saved themselves because they quickly become aware of the danger. This "early warning" is the best escape option since the victim got out in good shape while the fire was young and the structure intact. (A "rescue bonus" for Command.) These "do-it-yourselfers" will be either waiting outside, or in the process of leaving through the building exists, when the FGC arrives. They should be moved to an assembly area.

2. VICTIMS TRYING TO EXIT THE FIRE BUILDING. These victims will be having difficulty trying to save themselves and are usually in various degrees of panic (hanging over windowsills, jumping out of windows, etc.). Their condition may be marginal and their position precarious; therefore, they need immediate attention. They usually show a high willingness to be rescued. In some cases, the FGC must coordinate a fire attack with rescue functions to avoid "pushing the fire" onto these victims.

3. VICTIMS STILL INSIDE THE FIRE BUILDING (OR AFFECTED AREA). This group may be unaware, trapped, overcome, or somehow unable to save themselves. They are generally unknown to the FGC. He may initially estimate their number, location, and condition, but the only way he can accurately find these answers is by the completion of a primary search by rescue teams.

Searching For Victims

The search for victims takes place during the two major victim-finding activities, *the primary search and the secondary search.* A primary search initiated immediately upon arrival is not possible in all cases.

> **Primary Search:** A rapid search of all involved and exposed areas affected by the fire and which can be entered, to verify the removal and/or safety of all occupants
>
> **Secondary Search:** A thorough search of the interior fire area after initial fire control, ventilation, and interior lighting are completed.

The Primary Search

The FGC must structure his initial fireground operations around the completion of the primary search for all involved or exposed occupancies that can be entered. Primary search involves entering, locating, protecting, and removing.

There will be times when an interior search is impossible, as in the case of a fully involved structure. At this point, the FGC can initiate the search only when the structure is safe to enter. He must guard against futile, unsafe rescue attempts. If the structure is so well involved that it is unsafe for protected, well-trained personnel, then it is doubtful that any of the victims have a chance for survival.

REMEMBER: The safety of operating personnel must always be considered.

Often, rescue factors are not obvious. Victims are not always leaping out of the windows or lying on the windowsills. In these situations, there is a strong inclination to skip rescue and directly attack the fire. This may lead to the discovery of fire victims during overhaul, a grim reflection on the fireground management system.

Life safety is the absolute, number one fireground activity. The inescapable reality of fireground rescue is—the only way the FGC can be certain about occupant status is to send search teams into the involved and exposed areas to physically search for victims.

The critical factor duing primary search operations is time. The primary search must be fast and effective. The only way to guarantee standard search and rescue during every fire is to always conduct a search to verify occupant status on every offensive operation, whether or not actual fire is involved.

REMEMBER: The FGC must routinely extend a primary search in every offensive situation.

Processing decisions on the fireground takes time. A prefire decision (SOP) to make rescue the initial effort produces time savings where you need it most, at the beginning. The later rescue occurs, the less the chance of survival for the victims. Successful primary search operations must necessarily be extended quickly during initial offensive fire stages. Even though fire control activities may begin along with the primary search, everyone on the fireground must understand that this search has to be completed and reported before fire forces can move to the control priority.

FIGURE 4.1: Life safety is the number one activity.

During complex operations, the FGC must:

- Coordinate primary search assignments
- Secure completion reports from the interior
- Advise Alarm that the search is completed.

The FGC must be advised when the search is completed. He is the ONLY person authorized to send the appropriate radio transmission to Alarm. For the sake of convenience, the radio terms *"search and rescue"* should be used when ordering companies to perform a primary search and *"all clear"* to indicate that the search has been completed in a designated area.

When the FGC reports *"all clear,"* he is only reporting the completion of the primary search; he is not issuing a guarantee or signing an affidavit that everyone is out. The possibility of overlooking victims always exists. Sometimes victims hide from the firefighter as well as from the fire.

FIGURE 4.2: A primary search is not a guarantee that all victims have been found.

The primary search system is not absolutely foolproof. Extending a primary search means only that search teams have quickly gone through the interior to verify that everyone the search team can locate is out. The primary search is often done under hot, smokey, dark, rushed, and sometimes desperate conditions. Although the primary search system is not perfect, it offers the best chance of locating and removing victims.

The Secondary Search

The only way to absolutely confirm the presence or absence of victims is to make a secondary search after initial fire control operations are completed. If possible, this search should be done by companies not involved in the primary search, since primary search companies tend to recheck and verify their original search. In areas of major fire damage, this usually involves sifting through a lot of damaged property and demands manual labor. Thoroughness, rather than time, is critical here.

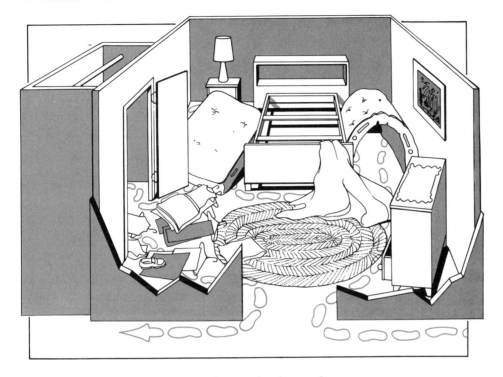

FIGURE 4.3: The secondary search must be thorough.

THE BASIC RESCUE SIZE-UP

CRITICAL FACTORS

There are four critical factors to be considered when developing a basic rescue size-up:

1. The fire stage
2. The fire victims—number, location, and condition
3. Effect of the fire on victims
4. Capability of control forces to enter the building, remove or protect the victims, and control the fire.

Command must quickly evaluate these factors and initiate operations, then continue to improve the quality of his life safety "intelligence" as the rescue proceeds. This basic rescue size-up becomes the framework for future rescue decisions.

Life safety is the most urgent reason for the FGC to call for additional alarms. He must, then, develop a realistic and pessimistic rescue size-up as early as possible and balance resources with the life safety problem.

The Fire Stage

The fire stage is a critical factor affecting the FGC's rescue approach. These stages and the corresponding appropriate fireground responses are:

> **"NOTHING SHOWING" or VERY MINOR FIRES (no life hazard)**—Initiate an interior search until it can be reported *"all clear."* Usually, it is unnecessary to remove occupants since *"all clear"* indicates they are in no immediate danger. The interior search also verifies fire status.
>
> **"SMOKE SHOWING" or "WORKING FIRE"**—Extend fire control simultaneously with rescue operations to gain entry and control interior access. The operation continues in a rescue mode until the primary search is completed and an *"all clear"* is transmitted. Follow this with a secondary search when the fire is stabilized.
>
> **"FULLY INVOLVED" BUILDINGS (or sections)**—Report status and that an *"all clear"* will not follow. Immediate entry and search activities are impossible and victim survival is improbable. Once the fire is under control, ventilate and light the fire area and begin a secondary search.

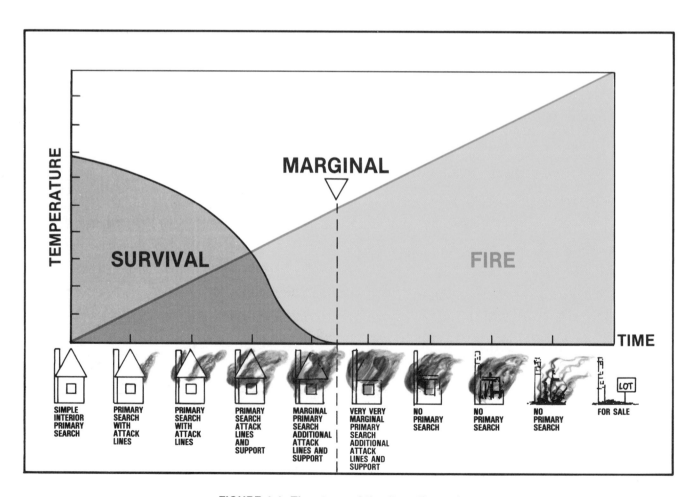

FIGURE 4.4: The stage of the fire affects the rescue approach.

The Fire Victims

Determining the number, location, and condition of the victims is not an easy task. There is an inclination for arriving units to ask spectators, "Is everyone out?" The problem with posing the question is in trusting the answer. The fireground quickly becomes confusing and chaotic, particularly during the initial fire stages, and it becomes a challenge to sort out all the people.

It is not functional to invest time interviewing spectators and then trying to determine their sanity and accuracy. Under the stress of fire conditions, even the occupants have difficulty producing an accurate role call. Keep in mind that some onlookers think that it is hilarious to yell "everybody's out" or "my baby's inside." Control forces should use reports on the location, number, and condition of victims as supporting evidence, but they must extend and complete a primary search whenever entry is possible. There is no other way to verify victim status.

Likewise, the location of a victim must be determined by direct search. Spectators may prove to be mistaken, or the victim, in an effort to escape the fire, may have moved from where he was last seen or would be expected to be located.

The condition of the victims may be predicted based upon the severity of the fire and smoke related to the structure. However, this is more of a guess than a true prediction. The time of involvement may not be too useful, since someone asleep may have been induced by smoke inhalation into respiratory arrest in a very short time. Generally, the longer the involvement and the more smoke and flame, the more severe the injuries to the victims and the more difficult the rescue.

Command must depend on interior search reports to accurately determine the number of rescuers needed, the difficulty of the rescue, and the type of care to be rendered. When safe to do so, interior crews should assess the victims using a First Responder patient survey and initiate basic life support. If conditions are life-threatening to the rescuer or to the victim, the SOP should insist on immediate removal of both from the danger area.

In addition to the physical condition of the victims, the FGC must also consider their emotional state. The more precarious the position of the victims, the more likely they will be difficult to rescue. Calm victims tend to ease the burden of rescue.

Effect Of The Fire Victim

The FGC must consider the current effect of the fire on the physical and emotional well-being of the victims and how this will change before and during rescue operations. This requires Command to size-up the fire, relate it to the structure involved, and predict its spread and the spread of smoke and gases prior to control. The order of search and rescue may be dependent on how this is done and success of the rescue may depend on how well it is done.

Control Forces

The activities of fireground personnel gaining entry and access, providing support activities, and control of the fire are covered in Chapter 8. Each factor must be considered when doing the rescue size-up.

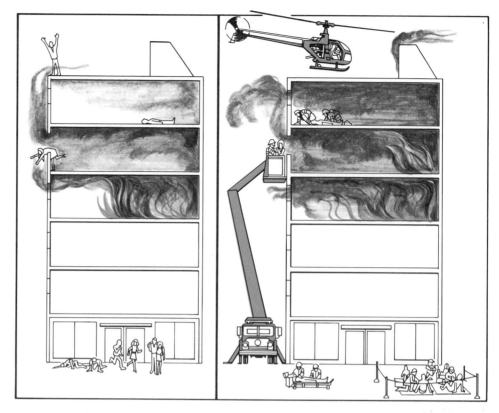

FIGURE 4.5: The possible effect of the fire on victims can determine the size and order of rescue operations.

The FGC must be familiar with the types of rescue activities, when they are called for, what equipment is normally used, what manpower is required for each procedure, what support activities are usually required, and how long it will take to perform various procedures. Such knowledge is essential in planning and carrying out rescue operations. When in doubt, the FGC should request estimates of manpower

FIGURE 4.6: The FGC must determine the types of rescue that may be needed and the time these activities require.

needs and time required for the rescue from his Rescue Officer or Sector Commander.

Unless the FGC has an understanding of rescue procedures, it will be impossible for him to make quick decisions regarding the rescue phase of fireground activities.

RESCUE OPERATIONS

THE RESCUE ORDER

Rescue efforts should be extended to victims in the following order:

1. Most severely threatened
2. Largest number (groups)
3. Remainder of the fire area
4. Exposed areas.

FIGURE 4.7: The rescue order serves as a structure to begin rescue activities.

The *rescue order* provides a structure for the initiation of rescue activities and the evaluation of resources needed, based on actual and potential rescue needs. However, even though it is a practical concept for fire personnel, fire victims may take a different view of the situation. Rescue operations include providing assurance for victims and taking other steps to reduce panic. The FGC must assign adequate manpower to manage all victims in the fire zones and exposed areas.

Keep in mind that certain handicapped victims may have to be moved higher in the rescue order. They may not be able to move to, or remain in, a safe area while awaiting assistance.

COMMAND DECISIONS

The FGC must ensure that certain activities are occurring, including providing safety for fireground personnel, quickly removing victims from life-threatening situations, properly caring for victims in safe locations, moving occupants away from the direct fire area, protecting victims from the elements, and preventing victims from re-entering the building.

All initial attack efforts must be directed toward supporting rescue functions. Hose line placement becomes critical. Attackers must control interior access (stairs, halls, lobbies) confine the fire, and protect avenues of escape. Everyone on the fireground needs to realize that the operation is in a *rescue mode* until the primary search has been completed. It may be necessary to hold the fire, and perhaps write-off the structure, in order to buy time for rescue.

FIGURE 4.8: Initial attack efforts must support search and rescue functions.

When dealing with large or complex occupancies, Command must make specific primary search assignments and must maintain ongoing control of these companies until the entire area is searched. This control includes providing replacement personnel when the primary search teams find and remove victims while other areas need to be searched.

Life Safety Decisions

As rescue operations are initiated, the FGC must make a basic, yet sometimes difficult, *life safety decision*. He must decide whether to:

- Remove the victims from the fire
- Remove the fire from the victims
- Use some combination of the two.

Like all fireground decisions, there are tradeoffs and risks associated with each option.

Fires which occur in residences and small commercial occupancies usually present a manageable offensive situation, i.e., a moderate number of potential victims along with adequate resources to handle the rescue. In these cases, the game plan is relatively simple—quickly implement a primary search, remove victims to a safe exterior position, extend a strong fire attack, complete a secondary search, and go home.

There will be other occasions when the FGC is faced with a fire or other emergency that has a large number of actual or potential victims. When the decision to evacuate the building has been made, his manpower pool must be drastically increased. It will take an army of firefighters and other rescue personnel to move large groups of people under fire conditions. His only salvation will be the rapid response of the resources necessary to rescue the victims who are still alive.

The Fireground Commander must very quickly develop an organization that will both stabilize the fire and provide for the removal of seriously threatened occupants.

While evacuation is common during most fires, there may be times when certain occupants are safer in their present location (rooms) than moving through contaminated hallways and other interior areas. Hospitals and high-rise buildings are good examples. Such movement may also impede or stop interior firefighting efforts.

A more effective approach is to move only those seriously threatened or affected and then begin an aggressive fire attack while leaving the remaining occupants where they are. The attack must be capable of protecting those occupants until the fire control is achieved. This concept requires that the FGC maintain an awareness of fire conditions while protecting the remaining occupants.

Controlling Victim Removal

Normal means of interior access should be utilized to remove victims whenever possible. Stairs, hallways, and interior public areas should provide the easiest and most comfortable exit for those inside. They also require less manpower to manage.

Secondary means of rescue include elevated platforms, aerial ladders, ground ladders, fire escapes, helicopters, etc. These devices should be used only when absolutely necessary and in their order of effectiveness.

> **REMEMBER: A rule of thumb for the FGC to consider is that the more he deviates from using the normal means of exit, the slower the rescue effort and the greater the resources required to do the job.**

Emergency Care

Another FGC concern is the treatment of victims after their removal. Victims should be moved to specified *triage areas;* one designated site is preferable. Scattering victims over the entire fireground requires more personnel. At this point, the proper use of basic life support (First Responders and EMTs) and advanced life support (Paramedics) personnel comes into play. Their capabilities are put to best use at this medical triage sector rather than wandering through the incident site. It may be necessary to assign rescue squad or engine companies as treatment companies to accomplish these medical needs within an acceptable time frame. Adequate Treatment Sector command is critical when treating multiple victims.

Stopping Re-Entry

Once the *"all clear"* has been transmitted, control of the fire area becomes a priority. A serious concern is always the occupants, family, or others who attempt to re-enter the building regardless of warnings. The most effective method to control the entire scene is to quickly establish fire lines, preferably with fire line tape or other highly visible means. A Police Liaison Sector established by the FGC will coordinate crowd and occupant control.

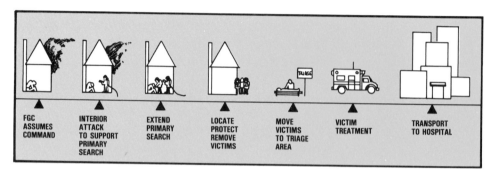

FIGURE 4.9: The FGC must understand all the steps involved in rescue operations.

SUMMARY

Rescue operations require fast, effective decisions by the FGC. Throughout the size-up and the rescue, *life safety* is the number one concern.

Rescue planning and operations are directly related to the *victim-to-fire position.* The closer the fire to the victim, the greater the danger. Victims may be outside on arrival, trying to exit the building, or still inside the fire building or affected area. A *primary search* is required to accurately determine the number, location, and condition of the victims.

A primary search is a rapid, but safe, search of all involved and exposed areas. All fireground operations must be structured around the primary search. A *secondary search* is a thorough search of the fire area after initial control, ventilation, and lighting.

The most important concern during a primary search is victim welfare and the safety of personnel. *Search and rescue must be safely conducted on every offensive operation.* Time is critical to the primary search since the later the search, the less chance of victim survival. Prefire planning lessens the complexity of search efforts.

The FGC must coordinate primary search assignments, secure reports from the interior, and advise Alarm of a completed search by issuing an *"all clear."* This is not an absolute guarantee that all victims are out of the fire area. That requires a secondary search.

Four critical factors govern a basic rescue size-up. They are the fire stage; the number, location, and condition of the victims; the effect of the fire on the victims; and the capability of control forces to enter the building, protect and remove the victims, and to control the fire.

There are three fire stages. *"Nothing showing"* usually does not require the removal of occupants. *"Smoke showing"* or *"Working fire"* requires the extension of both fire control and rescue operations simultaneously. *"Fully involved"* conditions make an initial primary search impossible. In all cases, once the fire is under control, initiate a secondary search.

The only way to be certain of the number, location, and condition of occupants in the fire area is to conduct a primary search with each offensive attack and a secondary search when the fire is under control. Use, but do NOT depend on, information gathered from bystanders.

The *rescue priority order* is: the most severely threatened first, the largest numbers next, then the remaining fire area, and, finally, the exposed areas.

During rescue operations, the FGC must decide if he is to remove the victims from the fire, the fire from the victims, or a combination of both.

All initial attack efforts must be directed toward supporting rescue functions. Most fires are manageable, offensive fires having a moderate number of victims and adequate resources on hand to carry out needed fireground operations.

The FGC should keep in mind that the more he deviates from normal means of interior egress and removal of victims, the slower the rescue effort and the greater the needed resources.

COMMAND DEVELOPMENT

To effectively plan and manage rescue operations, a FGC must understand the importance of the victim-to-fire location, the stages of a fire, the way a fire should develop for a given building or structure, basic rescue size-up, and rescue procedures. It is critical for the FGC to learn as much as he can about rescue procedures, personnel required, and the time for a given rescue activity. A FGC must learn how to work closely with Rescue Officers and personnel to obtain and understand the information he needs to size-up and conduct a rescue operation.

The following report card is provided so that you can evaluate your FGC knowledge and skills in classroom exercises, simulations, and on the fire scene.

Fireground Commander Report Card

Subject: Rescue

Did the Fireground Commander:

- ☐ Consider three categories of victims?
    - ☐ those already outside
    - ☐ those trying to get out
    - ☐ those still inside
- ☐ Extend a primary search in every offensive situation?
- ☐ Report completion of the primary search with an ''all clear'' radio report?
- ☐ Consider basic rescue size-up?
    - ☐ fire stage
    - ☐ number, location, condition of victims
    - ☐ fire effect on victims
    - ☐ capability of fire forces
- ☐ Structure initial attack to support rescue?
    - ☐ extend fire attack
    - ☐ gain entry
    - ☐ control interior access
    - ☐ confine fire
    - ☐ complete primary search
    - ☐ protect escape
- ☐ Make specific primary search assignments?
- ☐ Realistically evaluate victim removal requirements?
- ☐ Follow fire control with a secondary search?

5
FIRE CONTROL

MAJOR GOAL

TO EXTEND AN AGGRESSIVE, WELL-PLACED, AND ADEQUATE FIRE ATTACK.

OBJECTIVES By the end of this chapter, you should be able to:

1. State the major goal of fire control operations. (p. 144)
2. Define *"offensive"* and *"defensive* fire attack strategy."* (p. 144)
3. List the seven major factors and questions to be considered when distinguishing between offensive and defensive modes. (pp. 144-145)
4. State what type of attack is to be initiated when the mode is *marginal.* (p. 146)
5. Give an example of the actions taken by the Fireground Commander in response to mode changes. (pp. 146-147)
6. Distinguish between offensive, defensive, and marginal modes when given a list of fire scene descriptions, (pp. 144-148)
7. Describe, in seven steps, the basic offensive plan. (pp. 144-145)
8. Describe, in six steps, the basic defensive plan. (pp. 145-146)
9. List and describe the four basic variables to be managed by an effective attack plan. (p. 147)
10. Consider the most dangerous directions and avenues of fire spread and state how these factors affect fire control decisions. (p. 148)
11. State the most fundamental offensive fire control rule that regulates basic interior attack positions. (p. 149)
12. Describe how initial fire attack efforts are directed to supporting the *primary search.* (p. 149)
13. List the seven sides of a fire. (p. 149)
14. State the two basic, interrelated activities of fire control. (p. 150)
15. Explain why, in some cases, the most effective tactical analysis involves an evaluation of what is not burned, rather than what is. (p. 152)

DECISION MAKING IN FIRE CONTROL

COMMAND DECISION—OFFENSIVE/DEFENSIVE MODES

It is *standard operating procedure* for the Fireground Commander to extend an aggressive, well-placed and adequate interior attack wherever possible and to support that aggressive attack with whatever resources and actions are required to stop the fire extension and bring the fire under control. To achieve this goal, the FGC must consider, upon arrival and throughout the incident, whether the operation is to be conducted in an OFFENSIVE or DEFENSIVE mode. He should extend an offensive attack wherever and whenever conditions permit.

This is a *critical command decision,* leading to the selection of one of two strategies:

> **OFFENSIVE STRATEGY:** Interior attack, with related support, to quickly bring the fire under control.
>
> **DEFENSIVE STRATEGY:** Exterior attack, with related support, to stop forward progress and then to control the fire. The first defensive tactic is to protect the exposures.

The FGC decides if a situation is in the offensive or defensive mode through the analysis of seven fireground factors and their related characteristics.

This command decision is an ongoing one, requiring the FGC to reconsider these seven major factors throughout the attack. For example, the decision to begin, or continue a defensive attack mode, may be based on the fact that the offensive attack strategy has been abandoned (if it was ever begun) for reasons of personnel safety, and the involved structure has been conceded as lost.

COMMAND DECISION—THE OPERATIONAL PLAN

The FGC needs a simple operational plan for both the offensive and defensive modes. The following plans describe basic operating approaches that can be adapted and applied to individual tactical situations. Each plan is short and simple and offers the FGC a starting point and a framework for planning the attack.

Basic Offensive Plan

The purpose of the basic *offensive* plan is to allow for an *interior attack* that will confine and control the fire. This basic offensive strategy includes the following actions:

> **FGC ASSUMES COMMAND**—The command function must be continuous throughout the incident.
>
> **FAST, AGGRESSIVE, INTERIOR ATTACK**—First-arriving forces must immediately begin attack operations.
>
> **SUPPORT ACTIVITIES**—Quick development of resources

OFFENSIVE

1. FIRE EXTENT AND LOCATION
   HOW MUCH AND WHAT PART OF THE BUILDING IS INVOLVED?

2. FIRE EFFECT ON THE BUILDING
   WHAT ARE THE STRUCTURAL CONDITIONS?

3. SAVABLE OCCUPANTS
   IS THERE ANYONE ALIVE TO SAVE?

4. SAVABLE PROPERTY
   IS THERE ANY PROPERTY LEFT TO SAVE?

5. ENTRY AND TENABILITY
   CAN FIRE FORCES GET INTO BUILDING AND STAY IN?

6. VENTILATION PROFILE
   CAN TRUCK CREWS CONDUCT ROOF OPERATIONS?

7. RESOURCES
   ARE SUFFICIENT RESOURCES AVAILABLE FOR THE ATTACK?

DEFENSIVE

FIGURE 5.1: Seven decision points.

necessary to support the attack (ventilation, forcible entry, and provision of access).

PRIMARY SEARCH—Conduct a rapid, but thorough, search of the structure and surrounding area for victims, advising the FGC when the search is completed. Follow the basic rule: search every building, every time.

BACK UP INITIAL ATTACK—Provide support (back-up lines) for the initial attack crew, then cover the remaining exposure(s) (usually the rear).

WATER SUPPLY—Provide sufficient water for continued fire attack.

EVALUATE OPERATIONS/REACT—Quickly review attack success and modify strategy, if necessary.

OFFENSIVE STRATEGY

- FGC ASSUMES COMMAND
- FAST AGGRESSIVE INTERIOR ATTACK
- SUPPORT ACTIVITIES
- PRIMARY SEARCH
- BACK-UP INITIAL ATTACK
- WATER SUPPLY
- EVALUATE OPERATIONS/REACT

FIGURE 5.2: Steps of an offensive strategy.

Basic Defensive Plan

The purpose of the basic defensive plan is to allow for an exterior attack that will control the fire. A good defensive strategy includes:

FGC ASSUMES COMMAND—Again, the quick presence of "management."

EVALUATE FIRE SPREAD—Decide how much of the structure is a "write-off" because of fire spread before your arrival.

IDENTIFY KEY TACTICAL POSITIONS—Decide where to make the "stop" and position apparatus accordingly.

PRIORITIZE FIRE STREAMS—Decide the placement order and volume of attack lines necessary to make the "stop."
WATER SUPPLY—Provide sufficient water for continued fire attack.
EVALUATE OPERATIONS/REACT—Review attack success and modify strategy, if necessary. This generally means calling for additional resources.

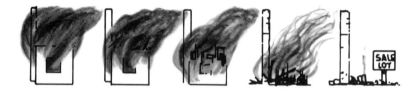

DEFENSIVE STRATEGY

- FGC ASSUMES COMMAND
- EVALUATE FIRE SPREAD
- IDENTIFY KEY TACTICAL POSITIONS
- PRIORITIZE FIRE STREAMS
- WATER SUPPLY
- EVALUATE OPERATIONS/REACT

FIGURE 5.3: The steps of a defensive strategy.

Marginal and Changing Modes

When the attack mode decision is clearcut, the FGC can quickly develop an attack plan; but when the situation is *marginal,* he may initiate a cautious, offensive attack while preparing, but not operating, defensive exterior positions. This strategy would be indicated where the FGC judges that an interior attack may not be ultimately effective but is necessary because of life safety considerations. An example would be a rapidly developing fire in an occupied building where the interior attack is needed to support rescue operations.

The FGC must be constantly sensitive to mode changes which may occur over minutes or hours. These potential changes require careful operations in marginal situations and an ongoing evaluation of the interior attack. He must:

- Coordinate the position and function of the interior crews with the Interior Sector Commander
- Manage information quickly and effectively

- Be highly responsive to changing conditions
- Not hesitate to order the Interior Sector out of the structure when the situation is deteriorating.

When a marginal interior situation is deteriorating, the FGC must be ready to pull out his troops and go to a defensive strategy. The control forces may object, making statements such as, "Give us another minute, Chief, and we'll put it out!" Such statements must not distract the FGC. The order to abandon the building means "everyone out . . . immediately." Only after the building is vacated, with crews reassembled and personnel accounted for, can the FGC reassess the situation. Prolonged evaluation of deteriorating conditions is too great a risk while personnel are engaged in an interior attack. The FGC must quickly evaluate the situation, but he must do so carefully in deteriorating situations.

In some cases, the order is given to abandon the building and the deteriorating situation does not fully materialize. Reassessment of the situation and reestablishment of the interior positions can take place smoothly. The FGC must be responsive to marginal situations without turning the attack into an all-night, in-and-out operation.

DEVELOPING AND MANAGING AN ATTACK PLAN

MANAGING BASIC VARIABLES

An effective attack can be developed by managing the following variables:

1. LOCATION/POSITION OF ATTACK—Evaluate options (offensive and defensive) provided by building openings (doors, windows, and arrangement of surrounding buildings).
2. SIZE OF ATTACK—Evaluate options of fire attack (manpower, hand lines, master streams, etc.) and translate into the size and number of hose lines.
3. SUPPORT FUNCTIONS—Evaluate the activities necessary to facilitate access and operations (forcible entry, ventilation, etc.) and integrate with other attack variables.
4. TIME OF ATTACK—Evaluate options of timing of fire attack (when to begin, duration, etc.).

These four major variables of where, when, how big, and what support are the standard hooks on which the FGC hangs attack decisions. These decisions determine the outcome of the attack. The list of variables is short. The challenge for the FGC is to correctly manage them. This must be viewed as a very special skill.

FIGURE 5.4: Four variables.

Planning An Effective Attack

The FGC cannot plan an attack or begin fire control operations until he determines *fire location* and *extent.* Sometimes locating the fire is simple; however, there will be situations where determining the location of the fire is a difficult task. Generally, the longer it takes to find a fire, the harder it is to put it out.

Decisions related to fire location and extent have to be ongoing. What is initially a nothing-showing scene can suddenly become a working fire. To properly plan and manage an effective attack, the FGC must:

1. Conduct whatever activities are required to first locate the fire and then determine its exact extent.
2. Be in a strategic position to react to reports of fire location and extent. This must be done as soon as possible upon arrival at the scene and continued throughout the event; while one group surveys, another must be prepared to attack.

It is crucial to establish and maintain control of fire search operations by initiating sectors quickly, making assignments for specific areas of the structure (quadrants) and managing this information effectively. Fire conditions can change quickly, so beware of allowing large numbers of firefighters to search the entire building unsupervised and situations where personnel have to be reassembled from throughout the neighborhood before firefighting can begin.

The FGC must now be prepared to react to these reports. Simply stated, when the fire is located, the FGC must be ready to attack it.

It is necessary for Command to develop an *attack plan* that takes into account the *seven sides* of a structure: the usual four plus the top, bottom, and interior. He must concentrate on the most dangerous direction and avenue of fire extension to stop fire spread.

Next, he considers the remaining sides and direction of travel, in order of danger, and begins action at those locations. Effective reconnaissance, early sectoring, and information control keep him updated on the entire fireground.

The FGC must consciously keep abreast of actions on every side of the fire, not just the most obvious side. He must rely on sectors, scouts, and reconnaissance agents for information about those areas he cannot see. If the availability of personnel is limited, the FGC may have to rely on one person to obtain and relay this information.

7 SIDES OF STRUCTURE
. 4 SIDES + TOP, BOTTOM
+ INTERIOR.

FIGURE 5.5: If you want to run around in circles, take up track!

FIRE CONTROL ACTIVITIES

APPLYING THE PLAN

Once the fire is located, the commander must structure the necessary operations to put water on the fire (put the wet stuff on the red stuff!). The rescue/fire control/extension/exposure problem will usually be solved by a fast, well-placed attack.

A powerful attack plan must be established and attack sectors managed to provide an aggressive attack that applies sufficient water *directly* onto the fire to overpower it. For most cases, the simple rule of **"small fire, small water; big fire, big water"** will apply.

Initial fire attack efforts must be directed toward *supporting the primary search.* The first attack must go between the victims and the fire to protect the avenues of escape. An attack from the interior, unburned side usually places inside forces in a position to accomplish this objective. The most effective escape path is normally through the unburned portion of the building. It is the FGC's responsibility to maintain control of the unburned portions of the building, particularly the stairs and hallways. There is a real temptation to go for the fire and neglect the primary search. The FGC must realize that he is in a *rescue mode* until he receives an *all clear* report, regardless of how much fire must be fought to get that report.

FIGURE 5.6: Small fire, small water; big fire, big water.

FIRE CONTROL PROCEDURES

Fire control involves two basic, interwoven activities:

STOPPING FORWARD PROGRESS OF THE FIRE (exposure-protection-confinement)
BRINGING THE FIRE UNDER CONTROL (extinguishment).

Often, in offensive attacks, both activities occur at the same time. The fire is attacked from the correct position (to stop forward progress) with sufficient force to extinguish it (under control). These tactics concentrate on *confinement* and typically occur inside structures.

When the size and intensity of the fire are *unmanageable* (defensive attack), the fire must be attacked from the appropriate direction, but with a *priority on protecting endangered exposures* and *stabilizing* (stopping) the *forward progress of the fire.* During this operation, the FGC should be assembling and managing additional resources until he has sufficient suppression forces to control the fire (e.g., an additional alarm).

During certain large fires, the Commander may only be able to stop the forward progress (keep unburned buildings from burning) until the fire burns itself down to a size that matches the capacity of the hose streams. This situation becomes, in effect, a controlled burning. Fireground activities then concentrate on *exposure protection* and are oriented toward *water volume*.

> *Offensive fires should be fought from the interior, unburned side* (basic offensive tactical factor). When fire forces do not have direction, they often will put water on the fire using the fastest, shortest route to the fire . . . ***"the candle-moth syndrome."*** However, this is not always the appropriate tactic. Attack from the burned side will often drive the fire, smoke, and heat back into the building and the interior forces out.

FIGURE 5.7: Offensive fires should be fought from the unburned side.

When fire is burning out of the building and not affecting life safety or exposures, let it continue to burn out while mounting an attack from the unburned side. It is probably venting in the proper direction.

When a fire involves *concealed spaces* (attics, ceiling areas, construction voids, etc.), it is crucial that these areas be opened up and fire streams be applied. Early identification and response to concealed-space fires may save the structure. Timing is very important— the officer must NOT hesitate to open-up in fear of damaging the building. The tradeoff is often no building a little later in the process.

All exposures, both immediate and anticipated, must be identified and covered. The first priority during defensive operations is *exposure protection.* The second priority will usually be to *knock down* the main body of fire. Although knockdown may assist in protecting exposures, it is not the first priority.

When the exposure is severe and the water supply limited, the most effective tactic is to put the water on the exposure. Once this coverage has been established, full attention can be directed toward knocking down the fire for the process of thermal-column cooling.

Should the FGC decide to change from an offensive to a defensive mode, the change should be announced as an emergency transmission ("emergency traffic"). All personnel must be withdrawn from the structure and held at a safe distance. During the retreat, interior lines will be withdrawn and repositioned, used as exit lifelines and abandoned if necessary. Lines should not be operated directly into doorways or windows but rather moved into positions to protect exposures and the personnel operating them. The FGC cannot waste time, energy, and resources on lost property. He must write-off involved areas and go on to protect savable locations.

As with an offensive attack, the best tactical analysis involves an evaluation of what is NOT burned, rather than what is or was burning. The unburned part represents where the fire is going and poses the problem for fire control.

After the fire companies have knocked the fire down, the Attack Sectors must determine if the fire is completely out. There is a danger of the FGC "letting his guard down" after completing what looks like fire control. This is a difficult fireground time. Crews are tired, the area is smoky, steamy, and often dark, and everyone wants to believe the fire is out. The FGC should not assume the fire is out and release companies prematurely. The standard radio reporting term *"Fire knocked down"* is a useful description of the fire attack progress that typically occurs before control is achieved.

The FGC is responsible for transmitting these reports to Alarm, which will record the time of the reports. *"Fire under control"* means the forward progress of the fire has been stopped, and the remaining fire can be extinguished with on-scene resources. It does not necessarily mean the fire is completely out but does indicate that the major hazards have been eliminated.

SUMMARY

A major goal of the FGC is to extend an aggressive, well-placed, and adequate *interior attack,* whenever possible. This attack is to be supported with whatever resources and actions are required to stop the extension of the fire and to bring the fire under control.

The strategy of attack is a command decision based on an *offensive* or *defensive mode.* The offensive strategy is an interior attack with the related support needed to bring the fire under control. The defensive strategy is an external attack with related support. Its purpose is to stop the forward progress of the fire and then control it.

The offensive/defensive mode classification is based on *fire extent* and *location,* fire effect on the structure, savable occupants, savable perty, entry and tenability, ventilation profile, and resources. Keep in mind that some situations are marginal at the beginning, while others may have changing modes. Therefore, all decisions involving the strategy of attack must be reviewed to allow for slow or sudden changes.

The basic offensive plan is structured to allow for an intensive attack that will *confine* and *control* the fire. The FGC takes command and launches a fast, aggressive interior attack. Support activities require the quick development of all resources necessary to clear the way for an effective attack. A primary search is conducted to cover the building and the surrounding area. Initial fire attack efforts must be directed toward supporting the primary search. The next step is the back-up of the initial attack. A key ongoing factor is to provide an adequate water supply for a continued fire attack. *Throughout the entire process of the offensive strategy, the FGC must evaluate operations and be prepared to match the attack to current conditions.*

An effective attack requires the management of four variables: location/position of attack, size of attack, support functions, and the timing of attack.

When planning an effective attack, the FGC must know fire location and extent. During the entire operation, he must be in a strategic position to react to reports. In order to establish control of the fire search operations, the FGC must quickly initiate sectors.

The attack plan must take into consideration the seven sides of a structure: *top, bottom, front, back, both sides,* and the *interior.* The plan must concentrate on the most dangerous direction and avenue of fire extension and provide a means to stop the fire in that direction. The remaining sides are then considered in order of danger.

Fire forces without direction may fall victim to the *"candle-moth syndrome."* A fast, well-placed attack is appropriate to put the necessary water on the fire. Good planning is necessary to overpower a fire.

Fire control involves two basic activities: *stopping the forward progress of the fire and extinguishment.* In an offensive attack, both activities occur at the same time.

If a fire is *unmanageable,* it must be attacked with priorities given to *protecting endangered exposures and stopping the forward progress of the fire.* Some large fires may require the FGC to concentrate activities on stopping the forward progress until the fire burns itself out.

Offensive fires should be fought from the *interior, unburned side,* preventing actions that could drive fire, smoke, and heat into the building.

When the fire is burning out of a building and it is not affecting life safety or exposures, it should be allowed to burn and the attack should be mounted from the unburned side. If the fire involves concealed spaces, these areas must be opened up and fire streams applied.

All exposures must be identified and covered. During a defensive operation, the first priority is exposure protection. The second priority is to knock down the main body of the fire.

COMMAND DEVELOPMENT

To properly plan and manage fire control situations, a FGC must continue to increase his understanding of the physics of structural fires. In addition to this, he must be thoroughly familiar with all fireground activities. As a FGC develops his command skills, he must concentrate on certain planning and managerial skills, including the analysis of fireground factors used to define the offensive/defensive mode, development of attack plans, quick evaluations of interior attacks, techniques to activate and manage attack plans, methods to allow for critical decisions relating to cutoff points, and methods to allow for effective forecasting to get forces ahead of the fire.

The following report card is provided so that you can evaluate your FGC knowledge and skills in classroom exercises, simulations, and on the fire scene.

Fireground Commander Report Card

Subject: Fire Control

Did the Fireground Commander:

- ☐ Consider standard factors to determine offensive/defensive mode?
- ☐ Extend a strong interior attack to confine and control in offensive cases?
- ☐ Protect exposures, stabilize forward fire progress, and surround and drown in defensive cases?
- ☐ Control position and function of control forces in marginal (offensive/defensive) cases?
- ☐ Consider most dangerous direction and avenue of fire spread?
- ☐ Attack from the unburned side? Resist "candle-moth" temptations?
- ☐ Structure initial attack to control interior access and to support the primary search?
- ☐ Apply adequate water directly on the fire?
- ☐ Consider all seven sides?
- ☐ Write-off lost property?
- ☐ Set up ahead of the fire and overpower it? Avoid playing "catch-up?"
- ☐ Open up and operate directly into concealed spaces?

6

PROPERTY CONSERVATION

MAJOR GOAL

TO KEEP PROPERTY LOSS TO A MINIMUM.

OBJECTIVES By the end of this chapter, you should be able to:

1. State the goal of property conservation. (p. 156)
2. List the four objectives to be considered during property conservation operations. (p. 156)
3. List the four steps involved in stabilizing loss. (p. 156)
4. Relate the timing of property conservation to the three major tactical priorities. (p. 156)
5. Define "primary" and "secondary" fire damage. (p. 157)
6. State how the Fireground Commander can communicate his concern for property conservation. (p. 158)
7. List the common property conservation tasks performed on the fireground. (pp. 158-159)
8. List four reasons for performing overhaul. (p. 159)
9. State how the Fireground Commander can improve scene preservation for fireground investigations. (p. 160)

BASIC PROPERTY CONSERVATION

OPERATIONAL OBJECTIVES

An important, yet often overlooked, function of the FGC on the fireground is property conservation. His responsibility is to commit the necessary resources required to keep property loss at an absolute minimum.

Effective property conservation activities require the same early and ongoing command functions and aggressive actions as the Rescue and Fire Control priorities. They generally produce more positive public reaction than any other fireground activity.

Property conservation activities generally come after rescue and fire control, but this is not an absolute rule. Some types of property (e.g., laboratories, libraries, and computer centers) contain equipment worth far more than the structure. In these cases, the priority may be to protect the contents first and to refrain from applying water in certain parts of the building.

Four objectives should be considered during property conservation operations:

- Stopping additional loss
- Verifying that the fire is completely extinguished
- Determining the fire cause and origin.
- Returning the occupancy to use, when possible.

To achieve these objectives, Command commits and directs companies in conservation activities that become the final stage of on-scene operations. The steps involved in stabilizing loss generally include:

- Evaluating damage to the overall fire area and the salvage of surviving property
- Deciding what conservation operations are required
- Committing the necessary personnel, equipment, and command
- Continuing to coordinate and manage conservation efforts until loss is stopped.

Considering that the FGC is responsible for every human being and physical thing on the fireground, he should manage his resources in a manner that protects life and retains, protects, and maximizes property value (a *"fireground conservationist"*).

During rescue operations, the FGC works to conserve lives; during fire control, he attempts to stop the fire; and during the third phase, he attempts to conserve property. When he has achieved the "all clear" and "under control" benchmarks for the rescue and control phases, he still must continue with property conservation responsibilities.

Timing

The timing of this "pure" property conservation stage is very important. Salvage operations cannot begin until the FGC is positive that the building is clear of victims. Likewise, there is little point in beginning salvage when the fire is still ravaging the building. If, however, the FGC identifies areas which are safe from the fire, but threatened by smoke or water, he may be wise to commit companies to begin salvage while the fire control efforts are still going on.

There will seldom be sufficient resources to begin all three priorities at once, so Command must usually follow a *1-2-3 sequence*. When Command reaches the stage when he can begin property conservation, he must commit whatever resources are necessary to keep property loss at an absolute minimum.

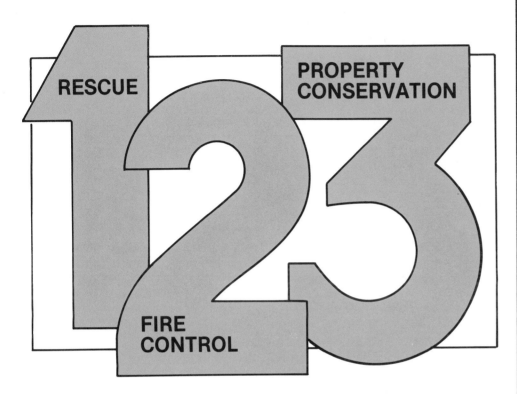

FIGURE 6.1: Command usually follows a 1-2-3 sequence.

To effectively conserve property, the FGC must understand the basic nature of fire damage and destruction. He must also correlate loss control with what is actually being damaged, how the loss is occurring, and when the damage is taking place.

Primary fire damage is caused by the basic products of combustion—flame, heat, and smoke. While the fire is producing these elements, the building and its occupants are being destroyed. Fire products affect both victims and property in similar ways, causing injury and death to one and damage and destruction to the other.

The best intervention strategy is to simply put the fire out as quickly as possible. After eliminating the fire, the FGC must deal with the residual effects of heat and smoke, or they will continue to do damage.

Secondary damage is caused by rescue, support, and fire control operations. Forcible entry, providing access, ventilation, checking for fire extension, and water application all do some damage. These activities are a firefighting damage tradeoff with the fire. In other words, the "FGC vs. the fire" game must be played in a way that saves as much as possible. The fire service has a unique relationship to all the property in the community as it is the only agency that can legally do damage during firefighting operations. The training, discipline, and actions of those on the fireground determine the final loss.

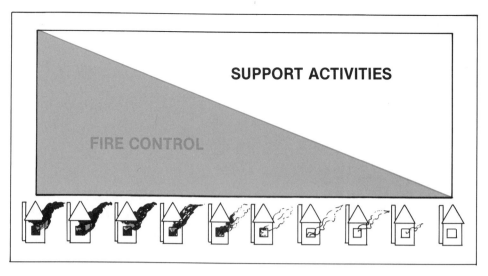

FIGURE 6.2: Disciplined support reduces fire damage.

Reducing Secondary Damage

Everyone on the fireground is expected to perform in a manner that continually reduces loss during operations. Unfortunately, some departments have the reputation of being highly destructive and showing little regard for property. These "demolition crews" will continue to face negative public reaction and a lack of support as long as they continue such actions.

The only way to reverse this image and create a positive one is to continually manage property in an effective, considerate, and professional manner.

The FGC should enforce responsible performance on the fireground and must not tolerate unnecessary damage to fire-stricken property. Firefighters cannot be allowed to expend energy, amuse themselves, or take out their hostilities by doing needless damage.

This concern and consideration for property is communicated to the firefighters by the effective management of all operations and a continual emphasis on loss reduction. The FGC becomes the pivotal property conservationist; *if it is important to him, it becomes important to his firefighters.* Almost every fire situation provides the opportunity to reinforce this professional image.

An important loss prevention benchmark is reached when the FGC turns his attention from fire control to property conservation. Firefighting operations require a considerable investment of time and effort from those involved. There is a natural inclination to continue control activities too long, particularly during offensive, interior attacks when nozzles tend to be operated until the furniture floats.

Often the FGC can initiate property conservation by recognizing that the forward progress of the fire has been stopped and ordering the lines shut down. This prompt shutdown stops further damage and signals the beginning of salvage work, freeing firefighters for salvage jobs. These hose lines will remain in place, however, until extinguishment and overhaul have been completed. *The earlier salvage begins, the smaller the loss.*

Some fires will allow Command to extend tactical priorities simultaneously, rather than sequentially, allowing for the overlay of fire control and property conservation efforts. Conservation resources need to be quickly evaluated. When the initial companies are actively involved in fire control (which they usually are) and salvage operations are left undone, additional companies must be called specifically for the salvage work. Early salvage is an important reason to quickly summon help to stabilize the loss.

The FGC should develop the following conservation mentality:

> **PLAY THE FIRE**—Consider how the fire is beating up the structure and damaging the contents. Build your attack plan around stopping the loss.
>
> **THINK LIKE WATER**—All matter is lazy and takes the easiest route. Send salvage crews to the lowest areas first to check for water damage (or to spread covers before the water arrives). Do whatever is necessary to stabilize hydraulic damage.
>
> **ACT AS IF YOU OWN OR LIVE IN THE PROPERTY YOURSELF**—Remove the most valuable items first, e.g., personal computers, stereos, etc.
>
> **SALVAGE EVERYTHING THAT CAN BE SAVED**—Evaluate the property profile and work to save anything that can be salvaged.

Property conservation tasks include making sure that the fire is completely out. Once the fire is under control, the FGC should structure a thorough survey of the entire fireground to evaluate extension and ensure extinguishment. Overhaul is important but not very glamorous. It stops additional damage from the original fire and eliminates the danger (and embarrassment) of rekindles. Overhaul also maximizes the value of the surviving property and returns the occupancy, when possible, to a safe and usable condition by cleaning up the residual effects of the fire.

FIGURE 6.3: It is senseless to carry out a complete overhaul of a total loss.

The FGC must beware of the premature releases of overhaul units. Holding a company a few minutes longer to finish the job may well prevent a second reponse later in the day for a rekindle incident.

Obviously, the amount of property that is savable will regulate conservation operations. Things that can be saved and occupancies that can be restored clearly deserve complete and effective overhaul. Since efforts occur at a traumatic time in the lives of the fire victims, they usually produce genuine goodwill.

On the other hand, there is little logic to the complete overhaul of a total loss that has burned to the ground. Senseless overhaul, which includes leaving neat piles of debris and clean floors, can only be attributed to "FGC psychosclerosis" (hardening of the head).

The FGC should maintain a sensible concern about the welfare of his personnel during long overhaul operations. Firefighters involved in rescue and fire control are generally fatigued, having reached their adrenaline "high" long before conservation assignments being. Ignoring fatigue may result in sloppy work and needless injuries.

Fresh companies should be brought in to relieve tired fire controllers when excessive overhaul is required. Rather than wear out a large number of firefighters by overhauling a large, total loss fire, Command should hold one company on the scene as a "fire watch" while releasing the others. Maintenance personnel, property owners, and managers hold the primary responsibility for janitorial service. The watch company can supervise the scene, handle small flare-ups, and alleviate the FGC's rekindle anxiety.

Scene Preservation

At the time fire control is achieved, the fire scene is as close to its prefire conditions (without reconstruction) as it will ever be. Command should be sensitive to this fragile state to preserve the scene for the investigator's examination and avoid creating the classic firefighter-investigator conflict. Firefighters want to shovel and go home; investigators want to sift through the remains to determine cause and origin. During these contests, the firefighters usually win since they outnumber the sleuths and already occupy the area. "Let's heave it faster than they can look at it."

Everyone on the fireground is responsible for assisting in the determination of fire cause and origin. All fireground operations should be accomplished with that in mind and all firefighters should continuously mentally log the conditions they encounter. The goal is to develop a total picture of the fire that traces back to its cause and origin. The perfect time to piece together this fire puzzle is immediately after the fire is placed under control. Even though the process of assisting in determination should begin at the outset, the period immediately after fire control is particularly critical.

An effective approach to integrating these two forces is to stop further firefighter operations after the fire has been controlled, pull firefighters out of the building, and let the investigators enter. In effect, the FGC replaces the Interior Sector with an Investigation Sector. This procedure gives firefighters a needed rest and interrupts shoveling of evidence.

Investigators will often request the use of one company to stand by with an overhaul line and lighting. The remaining units can return to quarters; if they are needed later for salvage and cleanup, they can simply return. Salvage can usually be postponed until after the investigators have cleared the scene.

SUMMARY

The FGC's attitude, methods, and willingness to commit necessary conservation resources are major factors in property conservation.

Property conservation must be viewed as a tactical priority, receiving the same command functions and aggressive actions as rescue and fire control. During property conservation, the FGC must set objectives to stop additional loss, verify that the fire is out, determine fire cause and origin, and, when possible, return the occupancy to use.

To stop additional loss requires evaluating the fire area in terms of what is lost and what is salvageable, deciding what conservation operations are necessary, and then committing and managing the necessary command, personnel, and equipment.

Timing of property conservation is critical. The building must first be cleared of victims and the fire must be under control. This means that command must follow the 1-2-3 sequence of tactical priorities.

Effective conservation of property requires the FGC to understand the nature of fire damage and destruction, correlate loss control with what is being damaged, note how loss is occurring, and be able to tell when damage is taking place.

The FGC must accept disciplined secondary damage (done for rescue and fire control) to reduce primary damage (fire damage). Property conservation must be important to the FGC and his personnel must be made aware of his feelings.

By recognizing when the forward progress of the fire has been stopped, the FGC can order lines shut down and reduce secondary damage. He should check for water damage in the lowest areas first.

His crew should know that the FGC believes in salvaging everything that can be saved.

Early salvage is important. For some fires, the tactical priorities can overlap, allowing additional companies to safely do salvage work during fire control operations.

During property conservation, the FGC must make certain that the fire is out. The FGC's anxiety over rekindles should be set to rest during property conservation.

Overhauls stops additional primary damage, maximizes the value of surviving property, and (when possible) returns the occupancy to use. Overhaul must be an effective operation. Why overhaul a total loss?

Firefighter fatigue is a key consideration during overhaul. Fresh companies are often needed.

Everyone on the fireground is responsible for assisting in the determination of fire cause and origin. The scene must be preserved for investigator examination.

COMMAND DEVELOPMENT

Tactical priorities identify three major tactical functions: rescue, fire control, and property conservation. These functions must be completed in a priority order to stabilize the overall fire situation. Having reached this point in the text, you should be able to list the functions (what to do), priorities (when to do them), and benchmarks (how to tell when each function is completed). The priorities are important tactical guidelines that provide a simple, straightforward framework for fire operations.

The following report card is provided so that you can evaluate your FGC knowledge and skills in classroom exercises, simulations, and on the fire scene.

Fireground Commander Report Card

Subject: Property Conservation

Did the Fireground Commander:

☐ Continually manage resources in a manner that minimized and reduced loss?

☐ Decide what property conservation operations would be required?

☐ Commit the necessary personnel, equipment, and command to property conservation?

☐ Continue to coordinate and manage property conservation efforts until loss was stopped?

☐ Time property conservation efforts with fire conditions and correlate loss control with what was actually being damaged?

☐ Insist on professional, responsible firefighting to reduce the secondary damage caused by rescue and control operations?

☐ Stop firefighting operations when fire control was achieved and order lines shut down in a timely manner?

☐ Overhaul completely to eliminate rekindles?

☐ Maintain a sensible concern about the welfare of his troops during long overhaul operations?

☐ Coordinate fire investigation activities with salvage and overhaul to assist in determining cause and origin?

7

FIRE STREAM MANAGEMENT

MAJOR GOAL

TO ACHIEVE TACTICAL SUCCESS THROUGH THE CORRECT SELECTION AND USE OF FIRE STREAM TYPE, SIZE, PLACEMENT, TIMING, AND SUPPLY.

OBJECTIVES By the end of this chapter, you should be able to:

1. State the goal of proper fire stream management. (p. 163)
2. List the five components of the tactical effect index. (p. 164)
3. Explain the correct decision process for applying hose streams to a fire. (p. 166)
4. Describe the characteristics of the following fire streams: solid, fog, 1½, 1¾, 2 and 2½ inch hand-lines, and master streams. (pp. 167-168)
5. State the primary function of hose lines in fire buildings. (p. 168)
6. List the four factors affecting hose line placement. (p. 169)
7. Describe the correct relationship between inside operations and external streams. (p. 173)
8. State the general rule of elevated fire stream use. (p. 174)

THE USE AND MANAGEMENT OF FIRE STREAMS

FIRE STREAM FACTORS

The FGC uses water as his basic weapon in almost every firefighting operation; it is the means by which he carries out his tactical plan. Water is the where, what, and when of that plan. The final outcome depends largely on the effective movement and application of water.

Fire stream management represents a major part of the overall fireground effort. Simply put, sufficient water applied directly to the fire will control the fire. The FGC must realize that at some point (the sooner the better) he must engage and fight the fire with water.

Everyone involved in firefighting activities should understand the characteristics, requirements, and hydraulic options to properly evaluate fire stream effectiveness throughout an operation.

Tactical success is an index of fire stream type, size, placement, timing, and supply.

Fire Stream Type

Firefighters are inclined to use the one type of nozzle that is most familiar, despite the great array of nozzles available. The clever firefighter will, however, expand his options by picking the correct nozzle for each specific job.

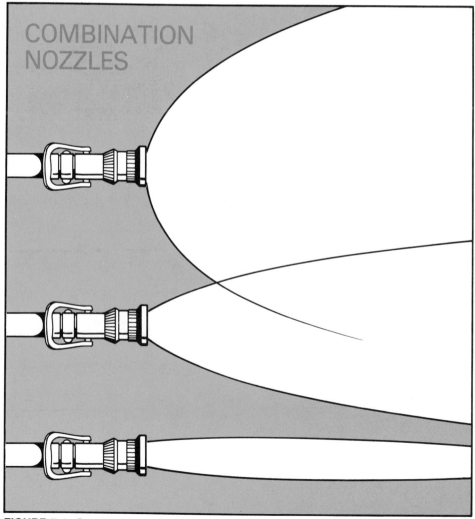

FIGURE 7.1: Combination nozzles provide a wide range of patterns.

In most departments, *combination nozzles* are usually attached to general purpose hand lines and work very well for that purpose. They provide a range of patterns, from wide angle to straight stream. They are highly versatile, and their fog patterns provide maximum heat absorption, expansion, and protection for the nozzleman in close-in or tight situations.

Fog patterns have a shorter range and less penetration capability on large volume fires than do straight stream patterns. By their very nature, they will entrain more air and sometimes can actually be used as exhaust fans.

Solid streams have the opposite characteristics—lots of power, reach, and penetration; less conversion to stream; and more ''quenching'' ability. While combination nozzles provide a straight stream setting, they cannot match the powerful solid jet of water produced by a straight bore nozzle tip.

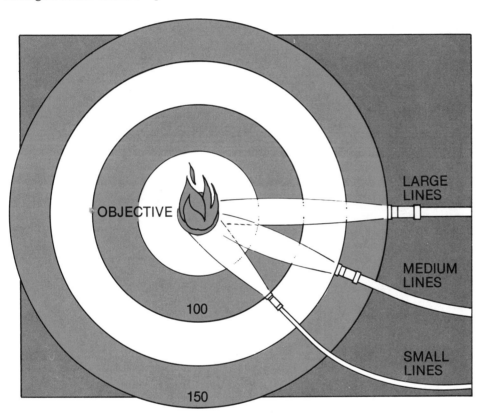

FIGURE 7.2: The line size decision is based on the maximum distance at which a line is effective.

When the size of the fire stream increases to larger hand lines and, sometimes, on to master stream devices, the use of the solid stream becomes a primary concern. Heavy duty tactics often require the striking power of smooth bore solid streams. Although solid streams sometimes are labeled ''old fashioned,'' they provide a modern and effective fire attack option.

REMEMBER: Use the correct type of nozzle for each job. Include solid streams along with fog. Heavy duty tactics with solid streams require having a solid stream line ready for immediate use.

FIGURE 7.3: An effective fire attack may require solid streams and fog.

Fire Stream Size

A critical fireground decision concerns how much water should be applied to the fire. Water must be applied at a rate sufficient to overpower the fire. This delivered effective volume is named the "rate of flow." Volumes below this rate will not extinguish the fire, while those

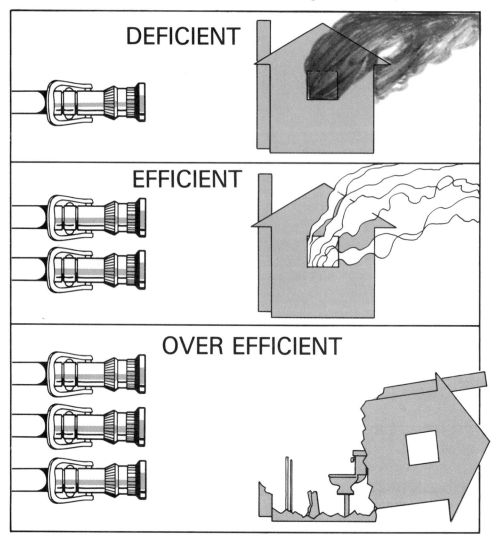

FIGURE 7.4: The rate of flow is dependent upon the size and number of hose lines.

above it tend to do excessive damage. The rate of flow is a result of the decision made when selecting the size and number of hose lines to be used and the nozzle type.

Choosing hose lines for firefighting is similar to choosing artillery for battle; the weapon must match the target. Guns range from rifles to cannons while fire streams range from booster lines to deluge guns. The smart firefighter will select and use, from the very beginning, the line size that eventually may be required. If a large line is needed, it should be pulled at the outset to avoid playing "catch-up."

FIGURE 7.5: You will have to play catch-up if you start with lines too small to fight the fire.

The characteristics of each type of hand line dictate the proper line choice. Each has a unique set of strengths and weaknesses, with a consistent tradeoff between speed, mobility, volume, and effort. Small attack lines are light, fast, mobile, and can be operated with fewer people. Larger lines are just the opposite. Obviously, the larger line provides a higher rate of flow.

Common hand lines may be grouped into three general categories—small, medium and large. Their characteristics include:

SMALL (BOOSTER LINES)—fast, mobile, and very low volume. They are supplied by tank water and are designed to be used on small, non-structural fires with limited potential, e.g., dumpster fires. Their convenience presents a temptation to use a booster when a larger line is required. Because of their low volume, any firefighting advantage is extremely marginal. Many departments have removed boosters from their engines altogether. (Basic Rule: If you pull a booster line, you better put the fire out!)

MEDIUM (1½, 1¾, AND 2 INCH LINES)—fast, mobile, medium volume. Provide excellent potential for aggressive, offensive firefighting for small to medium sized fires. They can be operated with limited personnel, offering the maximum gallons per minute (gpm) per firefighter. These lines are often used as a standard interior, fast-attack tool placed in preconnected, transverse hose beds. Sometimes, they can be overmatched during larger tactical situations. 1½ inch lines will only extinguish about 100 gpm worth of fire. 1¾ and 2 inch lines approach 2½ inch flow capacity with improved mobility.

LARGE (2½ INCH LINES)—slower, less mobile, large volume, with heavy knockdown, reach, and penetration capabilities. They require more muscle to use and move than do the smaller lines. They are very effective for larger, heavy-duty situations requiring big water. Even though 2½ inch lines are used less often than smaller lines, the FGC must be prepared to mobilize the manpower required to bring these lines into service.

TYPE OF LINE	SIZE	CHARACTERISTICS			
		FAST	MOBILE	VOLUME	REACH
SMALL	1" or less	YES	YES	NO	NO
MEDIUM	1½" to 2½"	YES	YES	NO	NO
LARGE	2½" or larger	NO	NO	YES	YES

FIGURE 7.6: The size of hose lines defines their use.

It is dangerous to use a marginal hose line, (one with just enough volume) which could be quickly overmatched. Fires can grow rapidly, requiring lines to have enough reserve to handle a dynamic fire increase while additional lines are being placed. It is better to have too much volume than too little.

Fires which have every available size of hose line on the ground usually indicate initial hose line misjudgment and mismanagement. This probably meant starting out with too small a line, then evolving through larger lines until master streams were employed. It is easy for firefighters to get into a habit of pulling the same line on every call since most fires in a given area are relatively similar (repeat business). The FGC must be prepared to order whatever size or combination of hose lines that are needed to accomplish the goal, particularly during non-routine fires.

Fire Stream Placement

Whenever possible, hose lines should be advanced inside fire buildings to control interior access to halls, stairways, and other vertical and horizontal channels through which people and fire can travel. Well-placed, adequate interior hand lines, including back-up lines should be extended quickly to maintain this interior control.

If the FGC loses inside operating positions and the ability to move within the building, the operation shifts to a *defensive mode.*

Initial hose line placement is regulated by the following principles:

- Place the first stream between the fire and any persons endangered by it. Protect the victims first and then protect their means of escape.
- When no life is endangered, place the first stream between the fire and the most severe (endangered) exposure.
- Place the second line to back up the first or to protect the secondary means of egress. Always consider the presence of personnel opposite this line.
- Place additional lines to support and reinforce attack positions in a manner and direction that assists rescue, supports confinement, and protects exposures.

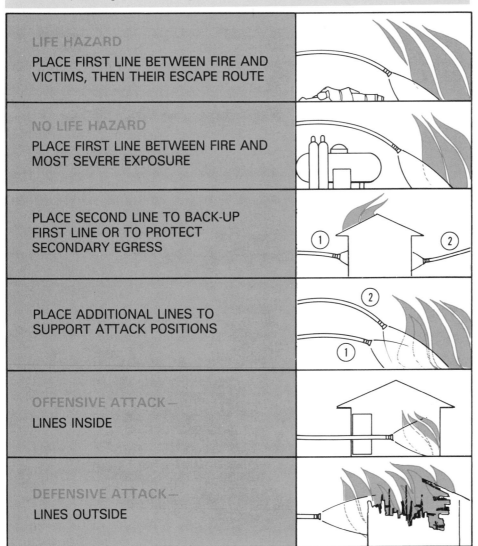

LIFE HAZARD
PLACE FIRST LINE BETWEEN FIRE AND VICTIMS, THEN THEIR ESCAPE ROUTE

NO LIFE HAZARD
PLACE FIRST LINE BETWEEN FIRE AND MOST SEVERE EXPOSURE

PLACE SECOND LINE TO BACK-UP FIRST LINE OR TO PROTECT SECONDARY EGRESS

PLACE ADDITIONAL LINES TO SUPPORT ATTACK POSITIONS

OFFENSIVE ATTACK—
LINES INSIDE

DEFENSIVE ATTACK—
LINES OUTSIDE

FIGURE 7.7: Initial line placement.

Offensive (interior) attack lines must be highly mobile and move in quickly through the unburned portions of the structure to the seat of the fire. Such operations can be described as aggressive, fast, active, vigorous, and forward. As the movement of these lines slows down, the operation becomes more defensive in both nature and effect. Most attack positions will not remain in the offensive mode for extended periods of time, so interior lines must impact the fire quickly. If water is applied to an offensive attack position and the fire does not go out, react. Either back up the position and put more water on the fire or move on to the next tactical position. Wherever possible, back up hose lines with the next larger size or with multiple lines.

Offensive attacks require fully protected firefighters (complete protective equipment and SCBA), both to enter the fire area and to stay in it long enough to engage the fire. One of the FGC's crucial functions is to decide whether the fire is to be offensive or defensive and then to control all the companies within that framework. The activities necessary to make the plan work are the firefighting tactics.

Firefighters have a tendency to set up hose lines at one tactical position and remain operating at this location for the remainder of the fire. The FGC should be sensitive to the real effectiveness of these lines after long periods of time. This "stationary hose line inclination" can result in fire streams being operated against blank masonry walls (firing squad streams), straight up (protecting the sky), going completely over the building (hydraulic arches), or shooting into smoke (penetrating the particles).

Since fire conditions change during the course of the fire, hose line operations must be constantly evaluated. This responsibility lies primarily with the Engine Company Officer.

This officer should be aware of where the stream is going and whether or not it is doing the job intended. Most often, the nozzle position is not the best place for this review. The officer must step back to get the whole picture. Sometimes this will require communication with other companies and sectors in adjacent positions to evaluate the entire impact of the attack. When a line is ineffective, the officer has several choices—move it, adjust it, redeploy it, or simply shut it down.

Fire Stream Timing

The effectiveness of fire streams is greatly increased when their operation is properly timed and coordinated with related fire functions. Attack lines should be in place and ready to go inside as soon as forcible entry is completed. If attack crews have to wait for water after the building is opened, additional loss will probably occur. Ventilation should also be timed with attack operations. Ideally, ventilation will occur just before the attack to clear the way for interior work. Venting too soon provides additional oxygen fueling fire growth; delayed ventilation adds to the punishment interior crews must face during the initial fire attack.

While moving inside the structure, interior lines should be properly placed and their use timed to reduce loss. Nozzles should not be opened until the fire is found. Do NOT shoot into smoke. *Fire control and loss reduction are achieved only when water is applied directly onto the fire.* Once the fire has been knocked down, nozzles should be shut down to reduce further water damage.

Fire Stream Supply

Another major factor in producing effective fire streams is water supply. The goal of this supply function is to provide sufficient water (agent) for firefighting as soon as possible. This requires the identification and utilization of an adequate water supply source and the delivery of that water to forward attack positions.

Specifically, this means a pump at the source and a supply line or tankers to move the water. Supply operations become much more effective when the efforts of multiple companies are coordinated. Speed, adequate supply, and maintenance of that supply require the collective efforts of more than one engine company.

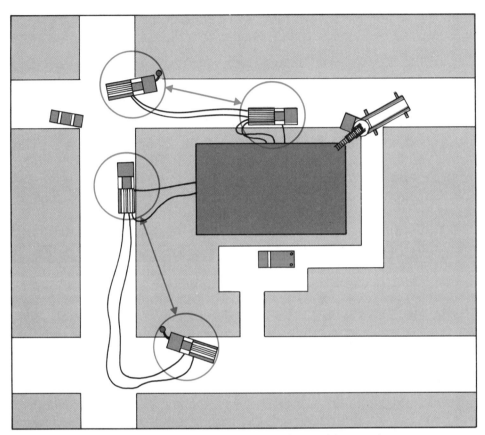

FIGURE 7.8: Supply operations require coordination and integration.

Water supply is based on the characteristics of the water source, whether it is an urban hydrant system or the drafting problem often found in rural areas. The attack capability depends on the amount of water pumped. When the FGC develops a strong water supply, he reinforces the entire operation and creates a built-in tactical reserve.

The Engine Company Officer is often faced with a wide variety of water supply challenges. He must find the best way to balance the water supply with the actual or potential size of the fire. His choices may range from making an initial attack from his booster tank (hoping for a quick knockdown) to requesting large diameter supply lines.

Regardless of the specific technique, each attack engine's goals must include providing its own uninterrupted and adequate water supply. This does not always include laying or pumping its own supply line. It is often more efficient to have another company perform these functions.

The classic Engine Company Officer's question, "Do I lay a supply line or go for the fast attack with tank water?" has no easy answer. Each choice has its tradeoffs. While laying a line provides a safe and adequate water supply, the fire is still burning. Using tank water is faster, but it is usually a marginal supply for a very limited time span. The better odds for the potential victims and firefighters lie in taking the extra time to provide supply lines from the initial approach. Additional incoming apparatus and other traffic can easily snarl or delay the crucial "supply line."

To make sound water supply and fire stream decisions, fire officers must constantly evaluate the current critical fireground factors. This process requires direct risk management on a scale from bold to cautious. Water supply operations are affected by the direct relationship between volume and time *(fast attack/small volume • slower attack/increased volume)*.

DIRECTION OF ATTACK

The direction of attack is crucial to good fire stream management, as is the effect of that direction on the fire. Everyone operating on the fireground must visualize the fire attack direction and movement in *"right-way"* and *"wrong-way"* terms. The right way is a smooth attack through the unburned portions of the building into the fire, pushing it upward and out through ventilation openings which are as close as possible to the seat of the fire. This takes advantage of the natural inclination of products of combustion to rise—a simple, yet important, reality of fire behavior. If the FGC chooses to ignore that reality, it can destroy his entire tactical plan.

Attack efforts should continually reinforce this correct direction by moving the fire away from surviving life and property. Any movement that backs the fire down to the uninvolved area or creates turbulence works against, and tends to reverse, the "right way" movement. "Wrong-way" movements can result from a burned side attack ("Candle-Moth Syndrome"), improper ventilation, opposing stream operation, and directing fire streams into vent openings.

Nozzle operation has a critical effect on fire movement in this rightway, wrong-way scenario. Attack lines move and drive the fire in ahead of themselves in the direction of their use. All hose lines, and fog streams in particular, displace much more air than water. The FGC must consider hose lines as fans that can be used to confine the fire and reduce the loss. He should regard nozzles as arrows that point in the direction where the fire will move when the nozzles are open.

FIGURE 7.9: Correct line placement and ventilation control the direction of fire movement.

Ventilation functions are another important part of correct attack direction. As the building is opened up, ventilation arrows are created that help to move combustion products up and out. The objective of correct ventilation is simply to move those products in a manner that supports the right-way direction. During offensive attacks, the arrows must be directed into the fire area to support primary search and to confine and control the fire. During defensive operations, the arrows are directed from outside the fire area to confine the fire and protect exposures.

When proper line placement and nozzle operation are combined with effective ventilation, the FGC gains the ability to manipulate the direction of fire movement. During the course of a fire, there are many temptations to violate the ''right-way'' attack arrows and reverse its direction. One uncoordinated hose line in the wrong place, discharging 5 percent of the total water being applied, can reverse the positive efforts of the other 95 percent.

The FGC must direct an initial attack that starts the arrows moving in the right direction and then provide the support necessary to reinforce that direction. He controls the arrows by developing an aggressive attack plan, building an effective organization by the establishment of sectors, and continually processing attack information.

He must then ''patrol'' the perimeter of that plan and, if someone violates it, redirect their efforts. The patrolling is not done by walking around the fire building but, rather, by carefully monitoring incoming information at the command post.

MASTER STREAMS

The placement and use of fire streams determines whether the overall operation is offensive or defensive. *Offensive lines go inside buildings, while defensive lines stay outside.* Inside and outside streams should not be operated at the same time or into the same fire area of the same structure.

When attack crews are working inside a building, exterior streams should not be directed into that building in any way which affects the safety or operations of the crews inside. Conversely, when exterior master streams are being used, all personnel should remain safely outside. **DO NOT COMBINE EXTERIOR AND INTERIOR ATTACKS.**

Master streams are typically used to blast large fires from the outside in defensive positions to cut off extension, confine the flames, protect exposures, and darken down the fire. They reach and penetrate fires that cannot be approached with hand lines. When you initiate an exterior stream, use a big one. Smooth bore tips deliver a large, more compact water stream.

Master stream appliances may be either portable or stationary (attached to apparatus). Portable deluge guns can be placed between, or sometimes inside, buildings and in locations inaccessable to apparatus. Once connected, they can be left unmanned, thus lessening the hazards to firefighters. Permanently mounted deluges (deck guns) can be quickly placed in service by the vehicle operator and are fast and convenient. They can also be plumbed directly into the pump, making for an even quicker attack.

Once master streams are charged they become basically stationary, as compared to hand lines, so their initial placement is the key to their effectiveness. Placement will generally be outside, between the things that are on fire and the things that are not (exposures). The object is to prevent the latter from becoming the former.

The placement and management of the larger diameter supply lines which feed these streams demands coordination; the Water Supply Officer must insure that the lines are adequately supplied. Master streams require a larger time investment than do hand lines, but their quick use to provide exposure barriers and heavy extinguishment will be well worth the time and effort. The FGC must order such streams into service quickly when encountering a large fire.

The FGC has several attack options using heavy streams. For example, he may choose to hold interior crews at a safe interior location while using exterior master streams to darken down the fire in a large structure. This requires a conscious FGC decision with close coordination and communication between sectors. Unfortunately, this simple rule is often ignored when firefighters are operating on their own with no overall plan or command. In these cases, the guy with the biggest line always wins and someone else usually gets injured.

ELEVATED FIRE STREAMS

The use of elevated master streams from ladder pipes, platforms, buckets, and booms can be both a blessing and a curse to the FGC. Their incorrect operation can create more problems than they solve.

Elevated streams are particularly useful for large, open fires involving piles of burning combustibles, exposure protection, and sweeping through outside windows during defensive operations. They are also useful when operating into well-involved buildings, generally where the fire has opened the roof. Because of their elevated position, these streams can easily affect the entire "wrong-way" fire movement and even reverse the rise of the products of combustion.

FIGURE 7.10: Elevated streams are used in the defensive attack mode.

To prevent incorrect use of elevated streams, the FGC must coordinate their placement, supply, and timing. *These streams are essentially a defensive fire stream,* so the FGC should have mentally "written off" a portion, or all, of the building before he initiates an elevated attack.

When offensive jobs are being done inside a vented fire building, the FGC must fight the inclination to put a cap on the fire by pointing elevated streams through the ventilation hole(s). Otherwise, both the safety of the firefighters inside and the remaining structure are in danger.

OFFENSIVE ELEVATED STREAMS	
DO	• EXTINGUISH PILES OF BURNING COMBUSTIBLES • PROTECT EXPOSURES (WET DOWN OUTSIDE SURFACES) • PROVIDE STREAMS THROUGH WINDOWS • PROVIDE STREAMS THROUGH BURNED, COLLAPSED ROOFS
DO NOT	• USE WITH PERSONNEL INSIDE STRUCTURE • AIM STREAMS THROUGH ROOF VENT HOLES • WASH AIR BETWEEN EXPOSURES

FIGURE 7.11: Operating offensive elevated streams.

Outside water application offers exposure protection, but it will not help the interior attack. Water applied to the outside of an intact roof will not affect the fire inside. The roof will shed water as it was designed to do and prevent water from reaching the seat of the fire.

If the fire has burned through the roof of a large building with a heavy fire load, the fire will probably be so overwhelming that even large, elevated streams will have minimal effect. Even as the roof begins to go, the corners and the underside of the collapse will be relatively untouched by elevated streams. At this point, the FGC uses ladder pipes to suppress the flames and then plays a waiting game. At some point, the fire will burn down to where the water flow rate equals the rate of heat release, allowing water application to extinguish it.

In some cases, the elevated streams may be only capable of cooling the thermal column sufficiently to reduce the danger of fire spread by radiant heat transfer or by intercepting flying brands.

When faced with a massive fire threatening to extend to the neighborhood of origin, the FGC must use water wisely. It should be directed onto the fire for cooling or onto the exposures to prevent ignition. Elevated streams projected gracefully into the space between the fire and the exposure can add to a photographer's "water festival" portfolio, but achieve little tactically. Generally speaking, the sky is noncombustible and, therefore, does not require protection.

SUMMARY

The FGC carries out his tactical plan using water as his basic weapon. Tactical success is an index of fire stream type, size, placement, timing, and supply.

The type of fire stream depends on the use of the proper nozzle. Since *combination nozzles* offer a wide range of patterns, they are very useful with general purpose hand lines.

Fog patterns have a short range and low knockdown capability; however, they provide maximum heat absorption, expansion, and protection for the firefighters using them.

Solid streams have lots of power, long reach, and good penetration. They provide less stream conversion and more "quenching" ability. Solid streams are most useful when employing larger hand lines or master streams, and can be used along with fog lines.

Water must be applied at a sufficient rate to overpower the fire without causing excessive damage. This delivered volume is called the rate of flow. It is highly dependent on the size and number of hose lines. Along with volume, the FGC must consider speed, mobility, and application effort when selecting hand lines. However, he must avoid selecting marginal lines but, rather, use whatever size or combination of hose lines that are needed to control the fire.

Hand lines should be used for close range, while master streams are used for long-range operations. Master streams blast large defensive fires from the outside in cases that do not permit the use of hand lines.

Master stream appliances can be either portable or stationary. They can be set up and left unmanned in dangerous locations. Once charged, master streams are basically stationary. Placement is usually outside, between what is burning and what is not.

If people are endangered, the first stream should be placed between the fire and the victims to protect their means of escape. When no life is endangered, the first stream should be operated between the fire and the most severely endangered exposure.

The second line is used to back up the first or to protect secondary egress. Additional lines support and reinforce attack positions.

Offensive lines go inside buildings; defensive lines stay outside. **DO NOT COMBINE EXTERIOR AND INTERIOR ATTACKS.**

The operation of fire streams must be properly timed. Ventilation must be properly timed with attack operations. Attack lines must be in place and ready to go inside once forcible entry is complete. Interior lines should be properly placed, with closed nozzles, until the fire is found and attacked.

Effective streams depend on an adequate water supply. Identification and proper utilization of this supply is critical. All operations must be integrated and coordinated. Remember, tank water is faster, but it is a limited and usually a marginal supply.

The direction of attack is critical for good fire stream management. A *"right-way"* attack is a smooth operation through the unburned portions of the structure and into the fire. This attack pushes the fire upward and out through the vent holes. The attack should move the fire away from life and property. The direction of attack is dependent on proper line placement, nozzle operation, and ventilation.

Elevated streams are useful for large, open fires. Improper use can lead to a "wrong-way" attack. These lines are mainly used during defensive modes. Outside application of elevated streams may endanger an interior attack.

COMMAND DEVELOPMENT

The FGC constantly matches the tradeoffs for a particular decision and should reinforce the same balanced approach by his officers. He must nudge the timid officer, who destroys more property by being too cautious, toward the bold end of the scale and harness the bold officer, who destroys more property by taking foolish chances, toward the cautious end.

Skillful fire control requires fast, well-placed streams with an adequate water supply which is coordinated, timed, and managed through the collective efforts of the FGC, Sector Commanders, and operating companies.

The following report card is provided so that you can evaluate your FGC knowledge and skills in classroom exercises, simulations, and on the fire scene.

Fireground Commander Report Card

Subject: Fire Stream Management

Did the Fireground Commander:

- ☐ Select fire streams large enough to control the fire?
- ☐ Utilize the proper nozzle?
- ☐ Use an adequate fire stream from the beginning—avoid playing "catch-up?"
- ☐ Consider the volume/time tradeoff?
- ☐ Place first lines between the victims and the fire and then between exposures and the fire?
- ☐ Control interior access with fire streams?
- ☐ Provide adequate back-up lines?
- ☐ Place fast, mobile streams inside buildings and large, powerful streams outside?
- ☐ Avoid combining offensive and defensive operations in the same fire area?
- ☐ Continually control a "right-way" attack direction?
- ☐ Evaluate fire stream effectiveness and change streams as required?
- ☐ Provide an adequate water supply for fire streams?

8

SUPPORT ACTIVITIES

OBJECTIVES By the end of this chapter, you should be able to:

1. Define and give examples for each major fireground support function. (p. 180)
2. Explain why timing is crucial to support functions. (p. 181)
3. Explain the importance of preplanning to the successful management of physical barriers on the fireground. (p. 183)
4. Explain the need for support functions on the fireground. (p. 183)
5. Define "primary" and "secondary" damage. (p. 188)
6. Explain the conservation approach to offensive operations. (p. 188)
7. Explain the relationship between the forcible entry effort and the level of security. (p. 189)
8. Describe forcible entry operations during "nothing showing" situations. (p. 189)
9. Describe the forcible entry "tradeoff" during "working fires." (p. 190)
10. List and describe the five major characteristics of effective ventilation. (p. 190)
11. List and define the four basic elements of ventilation. (p. 193)
12. List and describe the four major types of ventilation. (p. 194)
13. List at least seven rules for roof survival. (pp. 195-196)
14. Describe the importance of access operations to fire extinguishment. (p. 197)

THE NATURE OF SUPPORT ACTIVITIES

TYPES OF SUPPORT ACTIVITIES

Effective fireground operations require timely, well-placed *support activities.* These are the efforts that directly assist active fire rescue operations. Support activities open up and clear the way for interior operations. They also provide the capability for crews to gain entry and to remain inside long enough to operate directly on the fire. Examples of such activities include:

- Forcible entry
- Ventilation
- Providing access.

It is essential that the FGC be able to evaluate every tactical situation and coordinate the support efforts required for effective fireground operations. Each tactical situation requires a different level of support. The level of support is based on a number of critical factors, such as the nature and arrangement of the structure, the extent and location of the fire, and the need to operate inside with regard to the tactical priorities of rescue, fire control, and property conservation.

The balance between support and attack must be managed by the FGC through forecasting the need, position, and time for support.

FIGURE 8.1: If support activities are not planned and initiated at the correct time, prepare to conduct a long defensive incident.

When support is inadequate, poorly timed, or uncoordinated, forces cannot work directly on the fire and the operation will not be smooth and effective.

Often this breakdown occurs when interior crews encounter barriers that stop the forward progress of their attack. The increased confusion usually evident at this time should signal Command that progress has slowed, or stopped, and that the situation requires a quick tactical response. This alert response could include additional interior assistance in the form of more personnel, water, command, or attack positions.

When the response is late or ineffective, the scenario will rapidly become marginal. The FGC will have to consider changing to an overall defensive strategy—covering exposures and protecting personnel. Of course, some fires will demand a defensive approach from the start.

Fire control will be effective only to the extent that fire forces have direct access to the fire. Sometimes the fire can be reached easily, while at other times significant barriers will severely limit the ability to control the fire, regardless of the amount of water moved and applied by fire streams. Engine companies can produce enough water to fill up the building (bathtub tactics), but fire control still depends on the ability of support companies to do their job. The longer the fire hides in inaccessible places (behind locked doors, concealed spaces, attics, under unventilated roofs), the less are the chances of it being extinguished by control forces in time to achieve a positive outcome.

BARRIERS

During the initial stages of the fire, the FGC will have to focus quickly on those factors which present barriers to effective operations. These barriers delay the actions of firefighters and give the fire a chance to expand. Support activities, when performed correctly, manipulate these barriers to permit the firefighters to do battle directly with the fire and facilitate the achievement of interior suppression objectives. Timely support tasks open locked doors, provide ladders for access to above-ground positions, vent heat and smoke, and open concealed spaces that may reveal hidden flames.

There are two major classes of barriers:

1. Security
2. Construction.

Security

Most property is the scene of an unfortunately, ongoing war between two groups with opposite objectives—owners and occupants vs. robbers and burglars. This results in building openings that are protected with multiple locks, bars, bolts, chains, fences, barbed wire, booby traps, and psychopathic guard dogs. While the owners try to keep the bad guys out (stockade mentality), they also block needed access for the good guys. Firefighters must now be equipped and trained to force through the heavy-duty building armor to conduct interior fire operations.

Construction

The construction features of many structures incorporate barriers that inhibit direct firefighting. The fire hides in and around construction voids, attics, concealed areas, multiple ceilings, remodeled secret spaces, fake fronts, air conditioning shafts and ducts, open and unprotected channels, and anywhere else you cannot see into. Tough ceilings, walls, and unvented roofs cover these concealed spaces and usually resist ladder company tools. These obstacles require early identification and fast support to locate, confine, and extinguish the blaze.

FIGURE 8.2: Construction barriers often inhibit direct firefighting.

There is a third type of barrier which is often found when it is least expected. These are the barriers caused by vandalism or partial demolition. Sections of floor may be removed so that the firefighter enters a room and falls to the floor below. The same thing can happen with the removal of steps. Such hazards are certainly dangerous to personnel and may prove to be barriers to effective firefighting by adding firefighting personnel to the "victim" category. Sometimes vandals will cover such traps with cloth or construction materials with the intent to injure firefighters. They should always be anticipated in vacant buildings, on construction, renovation, or demolition sites.

PREFIRE PLANNING

Everyone in the fire business, including command guys, should expect to encounter fireground barriers as a matter of routine. The presence of physical obstacles should not come as a surprise to fire forces. Surprises cause delays, particularly during the first critical minutes of an attack. When an unexpected obstacle requires the commitment of support forces, the overall operation may become paralyzed.

The basic nature of these barriers and a standard approach to dealing with them must be considered during preplanning.

Prefire planning provides the attack team with advance information so they do not have to obtain it the toughest and most dangerous way possible—during a dark night while the fire is burning. The best time to determine that the doors are constructed from the deck of an old battleship or that there is a concealed space under the floor is during the prefire planning expedition—when conditions favor the firefighters. In addition to gathering information, the prefire planning process begins the decision-making function of command based upon a realistic expectation of the conditions that will actually be encountered.

TACTICAL POSITIONING

Tactical positioning is directly related to adequate support on the fireground. Such support allows the FGC to move companies into key interior positions from the unburned side. Usually, personnel approaching the fire from the unburned side will encounter more barriers than at any other position. Therefore, placing units in these spots will require more reinforcement than placing units in incorrect, exterior (easier), "candle-moth" locations.

FIGURE 8.3: Fireground tactics usually require approaching from the unburned side.

When tactical support is consistently lacking, companies begin to attack wherever they can (usually from the outside)—a bad firefighting habit during offensive scenarios. In contrast, when support is routinely extended as SOP, companies develop the inclination to position themselves in the correct interior spots, extend the primary search, and attack in a manner that maximizes effective rescue and property protection. When the FGC notices the attack crews are becoming "outstanding firefighters" (out standing in the street), it is time to re-evaluate their support. The lack of reinforcement could be the cause of this "open air" firefighting.

FIGURE 8.4: Lack of support activities leads to "open air" firefighting.

THE TACTICS OF SUPPORT

Support activities generally center on, and reinforce, interior offensive firefighting tactics. They must be extended during the period when offensive operations in and on the building can be accomplished. *Support jobs performed during marginal offensive/defensive situations can be very dangerous.* Management of firefighter safety is a primary concern for the FGC. As the fire moves into defensive stages, it will eliminate both the need and possibility for support. It will burn up locked doors, vent roofs and destroy attics, concealed spaces, and voids.

> **REMEMBER: As the fire begins to spread and affect more of the structure, the opportunity for conducting support operations will lessen.**

It should be obvious that support operations must be performed within a somewhat compressed time frame. This requires integration and timing between Command, Sector Officers, and forcible entry, ventilation, and the provision of access must be successfully combined with regular rescue and fire control.

Be aware that the outcomes of support functions are not objectives in themselves. They only contribute to the achievement of regular tactical objectives. Factor evaluation and reports from attack units support these regular priorities. Command must regard support as a standard element of company assignments and must be ready to plug in the necessary support resources to meet any tactical objective. The assignment of forcible entry, ventilation, and access provision provides this integration of support into the regular operating process.

Support Timing

Support timing is important. Ideally, support should go just ahead of rescue and firefighting to clear the way for forces to function directly in needed positions. Offensive operations can be visualized as attack arrows that penetrate the building in the correct direction to "kill"

FIGURE 8.5: To be effective, offensive firefighting requires adequate support.

the fire; support is provided along the way to create a smooth, fast, well-timed outcome. The FGC cannot wait for companies to get stuck in their assigned positions before providing support.

Virtually all support operations open the structure in some fashion. Premature support gives the fire an open-air head start before interior crews can rescue and attack. Likewise, delayed entry (forcible entry, access, and ventilation) gives the fire a chance to hide, burn, and expand. It also weakens the structure and subjects the interior crews to unnecessary danger.

Command and Sector Officers must be advised of the barriers encountered by attackers and of the specific support services needed. For example, "Engine 1 requesting ladder support to pull ceilings in the Interior Sector." Two-way communications is necessary so that enough support is at the right place at the right time.

Utilizing Personnel

Support functions typically are performed by ladder or truck companies and are usually referred to as "truck work." Squads may also get in on the action. Truck work is the very physical aspect of firefighting that involves breaking into spaces where the fire is burning (to do battle directly with it). Based on these requirements, truck company members ("truckies") have the reputation of being large, physical, and very active. They must be mechanically skillful and adept in the use of tools. Their activities are tactical and technique—oriented. Even though many jokes are made about truckies breaking a lot of things back at the station, these types of individuals are essential to aggressive interior firefighting.

An ideal division of work involves a coordinated combination of engine companies advancing hose lines while ladder companies clear the way and open up (provide support) just ahead of them. *This is a classic fire attack.* The FGC should create this "one-two punch" whenever possible to take advantage of the capabilities of each unit and the increased effectiveness that results when their efforts are combined.

Unfortunately, the Fireground Commander seldom lives in an ideal world and will not always have the right combination of units arriving at the right time. He may have to develop an attack plan with the units he has, while calling for the companies he needs. Ladder companies are the support specialists, but they are not always available. Many smaller departments simply do not have ladder companies, while in larger departments, because there are more engines than ladders, the engines usually arrive before them. Regardless, the fire is still burning and requires immediate action. The FGC may have to assign support jobs to another unit, such as the second-arriving engine. Failure to initiate support activities, opting instead to use the second engine to put more water on the fire, may lead to an offensive situation rapidly turning into a defensive one.

When ladder or squad companies are not available or they are late in arriving, the FGC should not hesitate to assign engine companies to perform specific support duties. The basic engine company should have the tools, training, and motivation to do a respectable share of opening up. A well-equipped ladder or squad should provide a better selection of power tools and other implements of destruction, but an ax and pry bar can open many doors and windows. In addition to the above items, all first-line pumpers carry at least one ladder that should allow an ambitious crew to safely make the second, or even the third, floor.

Every unit, particularly first-arriving engines, should be prepared to perform basic support functions to clear its own way for action. Engine companies cannot stop advancing the hose line because they encounter a locked door or other obstacle. They must necessarily use a "bash-as-you-go" program. This do-it-yourself approach works until a nonsupport company (e.g., primary search) encounters a major obstacle. When that happens, the FGC must be prepared to order heavy-duty support operations into action. It is crucial to identify the need and begin support as soon as possible. Many buildings have been lost because Command failed to recognize the need for support

FIGURE 8.6: Command must recognize the need for support.

or was reluctant to have the building opened up. Either hesitation causes the same disastrous outcome.

DAMAGE

Keep in mind the simple, yet important, guideline about damage—*primary damage is caused by the fire; secondary damage is caused by rescue and fire control.* Forcible entry, ventilation, providing access, and water application all do some secondary damage. The rule is—some secondary damage is necessary to reduce primary damage. The FGC accepts this fact when he first sends personnel into the building. He supports offensive operations in a building he plans to save. Based on that "save" approach, whenever truckies force a door, cut a roof, or make an access hole, they should select the option that is most effective, does the least damage, and will be easiest to fix when the fire is over. This *conservation mentality* should become a natural, standard philsophy on the fireground.

This train of thought does not naturally evolve on the fireground, particularly under difficult conditions. Companies will operate in a professional and responsible manner only after they have been trained under standard guidelines, procedures, and techniques.

Such preparation is reinforced when Command reviews actual field performance. This review requires the FGC to leave the command post after the fire to view support sites and to evaluate the effectiveness of firefighting operations (particularly access operations) with regard to timing, location, and loss. He also should consider occasionally making the walking tour while mop-up operations are still in progress, after assigning Command to another officer.

FIGURE 8.7: The FGC must evaluate the effectiveness of support activities.

Good performance should be commended, while screw-ups should be looped back into the training cycle. This "grand tour" sends a very practical message that loss is important when the FGC evaluates performance. If it is important to Command, it had better be important to the firefighter using an ax, hook, or saw.

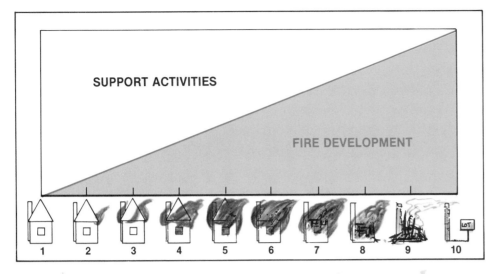

FIGURE 8.8: The FGC must evaluate support activity performance.

SPECIFIC SUPPORT ACTIVITIES

FORCIBLE ENTRY

When the firefighter encounters barriers that keep him from the fire area, forcible entry is required. Its degree of difficulty is directly related to the type of security encountered on the primary means of access (doors and windows). The FGC must attempt to match the forcible entry effort with the level of security. An imbalance, causes problems. *Excessive effort produces excessive secondary damage; minimal effort causes excessive delays and expands primary loss.* Initial fire conditions define the urgency for forcible entry. Conditions may vary from "nothing showing" to a "working fire," with each condition requiring a different forcible entry response.

"Nothing showing" situations allow the luxury of additional time to devote to a more delicate entry. This condition is evident when the

FIGURE 8.9: "Nothing showing" situations may allow for innovative entry techniques.

189

FGC notes no fire, no odor of smoke, and no occupants. The ever present "alarm malfunction" is suspected whenever a pessimistic evaluation reveals no hint of fire. It is irresponsible to do $4,500 worth of damage to a building that is "cold," particularly when management will be on the scene in five minutes with a passkey. "Nothing showing" scenarios allow the truckies time to try more delicate and innovative entry tricks, beginning with maneuvers that do the least damage.

On the other end of the scale, an active fire requires the sensible tradeoff between entry damage and the ensuing loss if the building continues to burn. Fast and effective entry is clearly indicated. Our "time vs. damage equation"—the faster you force, the more damage you do, gives way to the more critical the fire, the less important entry damage becomes. The luxury of time is gone; forget about picking the locks. Bash the barriers and go for a quick hit on the fire.

VENTILATION

Ventilation is a critical support function. Command improves the operation whenever he extends early and effective ventilation on active interior fires. Proper ventilation has the following characteristics:

- Prevents mushrooming
- Allows forces to gain and maintain entry
- Increases the safety of internal operations
- Improves interior visibility
- Controls heat and smoke damage.

Mushrooming

As interior fires burn, they quickly fill up the building with the products of combustion. As these products develop, they naturally rise,

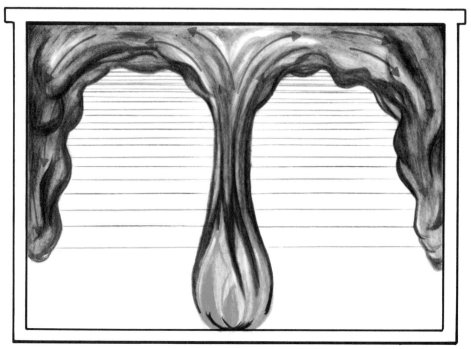

FIGURE 8.10: Mushrooming may occur with poor ventilation.

charge the top of the fire area, and then spread down laterally. This process is known as mushrooming. The mushrooming effect within an unvented building will fill the structure with smoke, superheated gases and, eventually, flame. This accumulation of unventilated combustion products is a critical factor that gives the fire the capability to take control of a major portion of the structure.

The phenomenon of *flashover* can occur with mushrooming. The gases trapped against the ceiling may suddenly ignite, quickly involving the whole interior space.

Mushrooming also sets the stage for *backdraft,* which is an explosion that may result from the sudden introduction of oxygen into a space-restricted fire. In tightly sealed buildings, gases will be confined, held in place by the ceiling or roof. When air is allowed to enter the confined spaces, these gases will ignite and explode. In most cases, the sign of possible backdraft will be grayish-yellow puffs of smoke escaping around windows and from under the eaves. Typical entry will allow the rapid influx of air and will probably cause a backdraft explosion. A structure with possible backdraft conditions will have to be ventilated at its highest point before firefighters try to enter the structure. More will be said about the dangers of backdraft later in this chapter.

FIGURE 8.11: The classical sign of a possible backdraft.

Mushrooming is a major fire-spread element and will eventually move an interior, offensive attack to the exterior, defensive end of the strategic scale. The combination of ventilation and rapid attack limits interior fire spread and usually will confine the fire to the original fire area.

To continue the offensive attack, the FGC must direct this vent/attack combination as far ahead of the fire build-up as possible. Ventilation opens up the structure, releases trapped heat, smoke and fire gases, and has the effect of ''burping'' the building.

Gaining and Maintaining Entry

Another factor in maintaining the offensive mode is the ability of firefighters to safely operate inside the fire building long enough to be effective. The unventilated build-up of combustion products will continue until the fire prevents firefighters from entering or drives them out of the building.

Correct ventilation lifts the smoke, heat, and fire gases and creates the tenability necessary for crews to finish rescue and attack the fire. Ventilating correctly provides visibility and improves the firefighters' abilities to control interior access. Often, correct ventilation makes the difference between offensive success and a building write-off.

Backdraft Prevention

The products of combustion are mostly flammable and present possible hazards to firefighters working in their proximity. An unvented fire is particularly dangerous because it quickly consumes the available interior oxygen and will eventually begin to asphyxiate itself. At this point of oxygen starvation, the fire usually has plenty of fuel left to burn. It also has generated a great amount of heat and only needs more oxygen to continue burning. The structure itself now serves as a barrier to the outside air and the building begins to display the classic "gasping" signs of a potential backdraft. The FGC cannot consider a tactical situation under control as long as a serious ventilation problem exists.

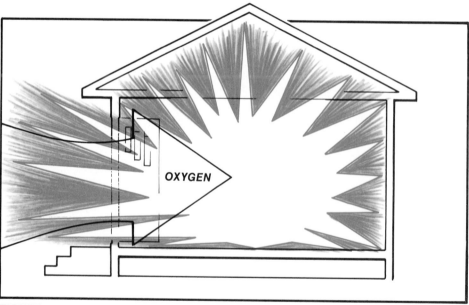

FIGURE 8.12: Backdraft can be created when oxygen is suddenly allowed to enter a confined fire.

Backdraft conditions create exciting and potentially fatal times on the fireground. If an entry is attempted under these conditions, an explosion, or backdraft, can occur. A backdraft is instantaneous and unforgiving and can seriously injure anyone in the general area.

The FGC's response to these conditions must be with fast, well-placed ventilation to relieve interior pressure and to begin channeling the fire to the outside. Early ventilation is the "explosion control" function used to reduce the exposure of firefighters to backdraft conditions.

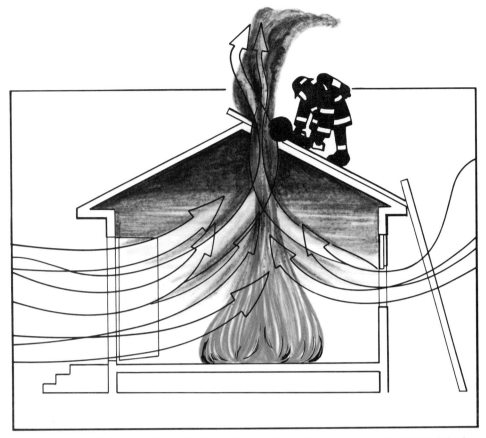

FIGURE 8.13: Well-placed ventilation will greatly reduce the hazards of backdraft.

Ventilation Tactics

Everyone on the fireground must understand how *ventilation supports the basic tactical priorities* and should have a standard plan in place to extend ventilation operations effectively and safely. Basic ventilation elements include the:

- Need to ventilate
- Timing of ventilation
- Type of ventilation needed
- Organization of support units.

Need To Ventilate

Virtually all interior fires require ventilation to remove the products of combustion from the fire area. Command and the Sector Officers must evaluate fire conditions on a scale from minor to serious and then select the appropriate type of ventilation. Fire conditions and ventilation are highly interrelated, and they must be balanced. In most cases, *minor fire—light duty ventilation; serious fire—heavy duty ventilation will apply.*

The FGC must avoid being controlled by fireground conditions. Venting is a major tactical tool used to manage circumstances and help the FGC control fireground conditions. It must be considered constantly and often becomes the deciding factor in successful offensive operations.

Timing Of Ventilation

Ventilation timing is extremely important and must be carefully coordinated with rescue and fire attack activities. Ideally, it should occur just ahead of interior crews advancing hose lines. Timing requires the FGC to evaluate (simply), to decide, and then to order the correct combination of actions. These steps must be done in the correct order and at the correct times to complete the current objective. He must mix and relate what, where, and when to produce effective fireground outcomes. Timing is the essence of command-control coordination.

Properly designed SOPs will allow the Roof Sector to act semi-independently of the FGC for certain venting operations. Ventilation timing needs to be coordinated between Roof and Interior Sector Officers. Well-trained crews should be able to coordinate timing and techniques and provide progress reports.

Type of Ventilation

The standard types of ventilation are:

- Horizontal
- Vertical
- Mechanical
- Hydraulic (water fog).

Each type of ventilation has its own unique characteristics. Choosing the correct type involves matching fire conditions with ventilation options.

Horizontal and mechanical ventilation utilize the regular configuration and openings of the building. They involve opening doors and windows and blowing out interior areas with mechanical fans once the fire is knocked down.

Vertical ventilation generally involves opening existing vents or cutting topside openings as close as possible to a point over the fire as safety allows. This type of ventilation is most effective for venting hot gases from a working interior fire. Vertical ventilation is very often the best way to control interior conditions during active offensive fires.

The ventilation effect created by the thermal expansion of water fog occurs when water is converted to steam, expands, and displaces the products of combustion. It becomes a side benefit when fog nozzles are used on attack lines. Water fog itself can also be used to move air through a door or window opening.

Organization Of Support Units

Correct ventilation requires assigning support units to the right position at the right time. Sector Officers also must be assigned to support activities to supervise, coordinate, and integrate operations with the overall attack plan. The FGC's ability to organize is directly related to his ability to evaluate fireground conditions. Information from the Sector Officers is absolutely necessary to the FGC.

During working fires, a great deal of ventilation activity may occur on the roof. Such activity requires a Sector Officer on the roof to act as a command partner for the FGC. He can perform a critical top-of-the-structure evaluation, gathering and forwarding data on roof conditions, fire location, extent, and travel. Likewise, he is in the best location to control operations for labor, effective work direction, and safety.

Ventilation is basically performed to alter interior conditions. Logically, then, the interior is the best place to determine if ventilation is required and, if so, what type and where. The Interior Sector can coordinate the location and timing of vertical openings with the Roof Sector. The two sectors working and communicating together greatly increase the overall effectiveness of venting operations.

The roof crews are in a position to do several things that can really help, or really hurt, interior operations. The Sector Commander on the roof must continually direct the truckies to resist the temptation to do "dumb" things above the fire. One common mistake is cutting vent holes in locations that spread the fire. In these cases, the fire will be channeled to burn out of the holes and will expand the loss. Vent holes should be located in a manner to support rescue activities and confine the fire.

Another set of problems relate to the inappropriate use of water above the fire. The "candle-moth syndrome" tends to overpower roof actors when fire and smoke appear from vent holes, even though this is exactly why you vent. The result of seeing this escaping fire and smoke may lead to hose lines and ladder pipes being stuffed down needed ventilation holes. The Roof Sector Officer must jealously guard such openings and provide the discipline required to control "roof stupidity." The guys on the roof should understand that hose lines above the fire exist for the purpose of protecting personnel and exposures.

Attic Fires

Most attic fires are controlled by pulling ceiling and applying hose streams from below. When attic fires occur with ceilings too high or too tough to penetrate from below, the FGC, as a conscious decision, may order a coordinated roof attack. This attack will include ventilation and suppression work from above and below the fire by enginemen and truckies.

REMEMBER: This is essentially a defensive attack on an offensive fire. It requires coordination, communications, command and control. The hose lines applied from the roof must be directed to confine the fire—not blast it down into the unburned area below.

In certain circumstances, a *"trench cut"* may be used to stop extension in an attic or under a flat roof. A trench cut is simply a manmade vent hole stretching from the roof peak to the gutter or completely across a flat roof. A trench cut is designed to be a line the fire cannot cross and is extremely effective if it is protected by hose lines and is made before the fire reaches its location.

Ventilation Safety

Tactical positions above a fire are very dangerous spots; firefighter safety must always be a primary consideration during ventilation operations. Remember to approach truss and wide-span roofs with extreme caution. All roof personel must stay alert to the rules of roof survival, and be aware that these rules are based on both fire and structural conditions. These rules include:

- "Read the roof" before the fire—some types are very dangerous, e.g., paneled roofs
- Probe ahead with a tool before stepping on to the roof
- Establish the initial position in the safest area

- Use this "safe area" for possible retreat or refuge
- Work in pairs
- Maintain multiple escape routes (at least two ladders)
- Check for weakness before walking onto an area
- Constantly check roof conditions
- Walk only on support members
- Utilize small "inspection holes" before beginning large cuts
- Keep the number of roof firefighters to the absolute effective minimum
- Use SCBA when above the fire
- Leave the roof when the job is done.

Everyone working on the roof should be wearing SCBA. Conditions often change very rapidly and this air supply may prove to be a lifesaver. SCBA must be used any time there are firefighters over an active fire area.

Another common, yet serious, problem is simply that the roof group stays on the roof too long. There's no tactical advantage in having the truckies watch the products of combustion escape from the hole they just cut. Once ventilation jobs are complete and there are no serious exposures present, it is time for the Sector Officer to get the "shepherds" off the roof.

FIGURE 8.14: When venting jobs are completed, get the shepherds off the roof.

Roof conditions are an important indicator of an offensive/defensive attack mode. If the truckies cannot get onto the roof, or they get chased off because of advanced fire conditions, then it is time to seriously consider a shift to defense. This is true even if the interior crews can still stay inside. The interior guys will be upset if they end up wearing the roof. Interior information is not enough. The FGC must have accurate, up-to-the-minute roof information. Consider a "roof clear" radio notification similar to the search and rescue "all clear" used earlier.

ACCESS OPERATIONS

Buildings typically are constructed so that virtually everything inside or behind the interior finish is a hollow concealed space (inside walls, floors, ceilings, attics, etc.). As fires burn, they beat up the interior finish and eventually will burn into these concealed areas. Fires that originate inside these spaces are hidden and create special problems because they can rapidly expand before discovery. When concealed construction voids are present, they are usually widespread and expose a major portion of the structure. Because of that interior exposure, concealed spaces can be bad news to the FGC; if they are not discovered quickly and controlled quickly, they can do major, if not total, damage.

The basic challenge for truckies faced with concealed space fires is to open up the area to allow water to be applied directly onto the fire. *The ability to gain access usually determines whether or not the fire is extinguished.* If the involved void spaces are not opened up, the fire will have a chance to extend and become deep-seated throughout major portions of the structure. When the fire involves, consumes, and moves out of the concealed area, it will usually expand to the point where the entire operation is catapulted into a defensive mode.

Access-oriented jobs involve pulling ceilings, opening walls and floors, and the support activities required to allow attack on the hidden fire. To be effective, these jobs must be timely and well placed. Command and Interior Sectors cannot be timid during the size-up of a concealed space fire.

Initial hesitation to open up will usually force everyone to chase the fire. Access, as a form of support, requires good size-up to be at the right spot, quick decisions to be on time, hard work to get inside, and good coordination to balance access with attack. The FGC must:

- Forecast fire travel
- Get firefighters ahead of the fire
- Open up the structure
- Cut off the fire
- Complete extinguishment.

Command also must continue to provide the necessary support to open up even after the fire is knocked down.

When initial fire control is achieved, or thought to be achieved, the FGC should assume a pessimistic appraisal of whether or not the fire is extinguished. Access management requires "FGC flexibility"—bold at the beginning, cautious at the end. During this period, the experienced FGC holds the companies a bit longer to completely open up all concealed spaces and to double check that the fire is definitely out. "Checking for extension" is an essential step to verify that the fire has not found any good places to hide from the attack.

SUMMARY

Support activities are an essential part of effective fireground operations. Such activities include forcible entry, ventilation, and providing access. Various factors determine the level of support, including the nature of the structure, the extent and location of the fire, property conservation, and the phase of operations (rescue, control, etc.).

The FGC manages the balance between support and attack by forecasting the need, position, and timing of support. Poor management of support activities can change an offensive incident into a defensive one.

Fire control requires forces to have direct access to the fire. The FGC must focus support activities on removing barriers that delay the actions of firefighters. These barriers can be due to security or construction.

Support activities generally center on and provide reinforcement for interior firefighting. Therefore, *support should be seen as an offensive situation activity.* As a fire spreads, the opportunity for support activities decreases and the risk to personnel increases.

Preplanning is the key to having the needed information required to make timely fireground decisions concerning support.

The outcomes of support functions are not objectives. They only contribute to the tactical objectives. Command must view support activities as a standard element of company assignments.

Tactical positioning is a major factor in providing adequate support. If key interior positions are taken from the unburned side, the crews will encounter more barriers, thus requiring more reinforcement. Support, extended as an SOP, helps companies develop the inclination to properly position themselves.

Support should go just ahead of rescue and firefighting. Command and Sector Officers must be advised of barriers, fully utilizing two-way communications to correctly time support activities.

Premature support encourages fire spread. Delayed support gives the fire a chance to hide, burn, and spread.

Support functions are usually performed by ladder (truck) companies. They clear the way and provide support just ahead of engine companies. When ladder companies are not available, the FGC should assign engine companies to support duties, resisting the temptation to delay support in favor of putting more water on the fire. Every unit must be prepared to provide support.

Primary damage is caused by the fire; *secondary damage* is caused by rescue and support operations. Conservation of property requires controlled secondary damage. Effective fire control requires some secondary damage to reduce primary damage.

One way to reduce unnecessary secondary damage is for the FGC to evaluate this damage in regard to timing, location, and loss.

Forcible entry is required when a barrier prevents access to the fire area. This entry should be matched with the level of security. "Nothing showing" situations allow time for innovative entry techniques. An active fire justifies more damage—the more critical the fire, the less important the amount of entry damage.

Ventilation is a critical support function that prevents mushrooming, allows forces to gain and maintain entry, and increases the safety of internal operations. A vent/attack combination has the effect of "burping'' the building, reducing the threat of mushrooming, flashover, and backdraft.

Correct ventilation will lift smoke, heat, and gases to support entry, visibility, and position for interior activities. Venting improves the firefighter's ability to control interior access.

The basic ventilation elements include need, timing, type of ventilation (horizontal, vertical, mechanical, and hydraulic), and organization for assigning the support units to the right position at the right time.

A great deal of ventilation activity occurs on the roof, requiring a Roof Sector Officer and close coordination with an Interior Sector Officer. The FGC must be certain that vent holes are not cut in locations that will spread the fire or be a danger to interior crews. All firefighters must be prevented from using the vent holes to add water to the fire or to provide access for hose lines and ladder pipes.

Ventilation safety requires the establishment of initial positions in the safest area, continued use of the safe areas for retreat and refuge, multiple escape routes, constant checking of roof conditions, keeping personnel on the roof to an absolute minimum, using SCBA when over an active fire, and leaving the roof when the venting job is done.

Access operations are based on the fact that everything inside of, or behind, a structure's interior finish is a hollow or concealed space. These areas must be opened up to allow water to be directly applied to the fire. The FGC must forecast fire travel, get firefighters ahead of the fire, open up, cut if off and put it out. The FGC must be bold at the beginning of access operations and cautious at the end. He must assume a pessimistic appraisal of fire extinguishment.

COMMAND DEVELOPMENT

Effective fire command requires the FGC to understand support activities and how they are an essential part of fireground operations. The FGC must develop a working knowledge of construction and how fire spreads in various structures. The FGC must learn to utilize the information gained prior to a fire to develop practical firefighting tactics that include probable support activities.

The following report card is provided so that you can evaluate your FGC knowledge and skills in classroom exercises, simulations, and on the fire scene.

Fireground Commander Report Card

Subject: Support Activities

Did the Fireground Commander:

☐ Preplan and evaluate barriers?
☐ Forecast need and position of support?
☐ Utilize forcible entry, ventilation, and access operations to place companies in correct positions?
☐ Time and coordinate support with attack?
☐ Control support operation damage?
☐ Provide adequate forcible entry just ahead of attack line entry?
☐ Provide effective ventilation to:
   ☐ prevent mushrooming?
   ☐ gain and maintain entry?
   ☐ assure firefighter safety?
☐ Quickly open up concealed spaces?

9

APPARATUS PLACEMENT

OBJECTIVES By the end of this chapter, you should be able to:

1. State the major goal of apparatus placement. (p. 201)
2. List the three basic objectives of apparatus placement. (p. 202)
3. List the four factors affecting apparatus placement on the fireground. (p. 203)
4. Cite the five categories of apparatus on the fireground. (pp. 203-204)
5. Explain the term "key position" as it relates to the fireground. (p. 208)
6. Describe key positioning on the fireground for engine companies (p. 208), for ladder companies (p. 210), and for command vehicles (p. 212).
7. Explain the need for relay pumping operations. (p. 209)
8. Define the term "key hydrant." (p. 210)
9. List the minimum distance that apparatus should be positioned from any structure on the fireground. (p. 215)
10. Describe the correct positioning of aerial apparatus during non-rescue situations. (p. 211)
11. Define the term "attack team." (p. 213)
12. Explain the usual deployment of an attack team on the fireground. (p. 213)
13. List and define the four standard stages of operations. (pp. 213-214)

GENERAL DEPLOYMENT

OBJECTIVES OF APPARATUS PLACEMENT

Virtually all fireground operations involve fire apparatus in some way. The location and general deployment of that apparatus will, to a major extent, regulate how effective the equipment will be and how well the Fireground Commander can utilize the company within his plan. The success of most fireground operations depends directly or indirectly on the effective use of fire apparatus.

There is a direct relationship between apparatus placement and function. Certainly, there will be situations where this rule must be modified, but, whenever possible, apparatus should be placed where it can be used most effectively. Unfortunately, we often reverse this basic rule, thus, limiting, or even eliminating, the functions that could be performed by a particular unit.

REMEMBER: Correct apparatus placement strengthens the overall fireground operation system by expanding the capabilities of all companies on the scene.

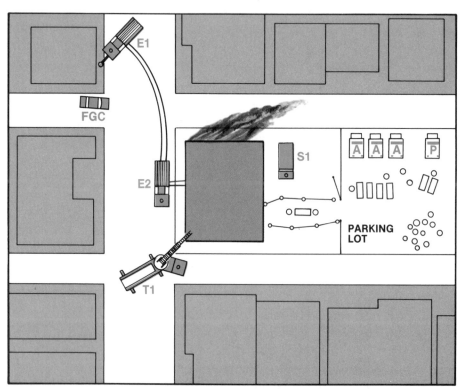

FIGURE 9.1: Apparatus function should regulate placement.

The basic objectives of apparatus placement are to:

- Place every piece of apparatus to take maximum advantage of its capabilities
- Utilize only the companies that are required
- Expand the FGC's options with a tactical reserve of uncommitted companies in standby positions (staging areas).

DECISION ITEMS

Firefighters have a natural inclination to drive fire apparatus as close to the fire as possible, but this often results in positions that are dysfunc-

202

tional and dangerous. Unless controlled, there will be a tendency to have all responding apparatus driven directly to the fire. First arrivers should take *key forward positions*, while later-arriving companies should take *staged positions*. Staged companies should stop short of the immediate fire area and should remain uncommitted until ordered into action by the FGC. Company Officers should select staged positions that afford maximum tactical options.

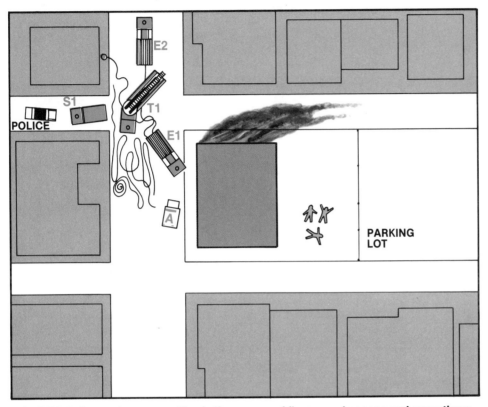

FIGURE 9.2: Apparatus congestion is the enemy of fireground access and operations.

Apparatus placement should hinge on one or more of the following:

- Standard Operating Procedure (SOP) for first-arriving companies
- Pre-arranged staging procedure
- Conscious decision of Company Officer, based on existing or predictable conditions
- Direct order from the FGC.

Throughout fireground operations the FGC must be aware that apparatus access controls his tactical options. Unless great care is exercised, the immediate incident area can quickly become congested with fire units, reducing those options drastically.

SPECIFIC DEPLOYMENT

THE BASIC CATEGORIES OF APPARATUS PLACEMENT

Apparatus on the fireground falls into one of five basic categories:

1. Responding
2. Staged
3. Operating
4. Parked
5. Returning to quarters.

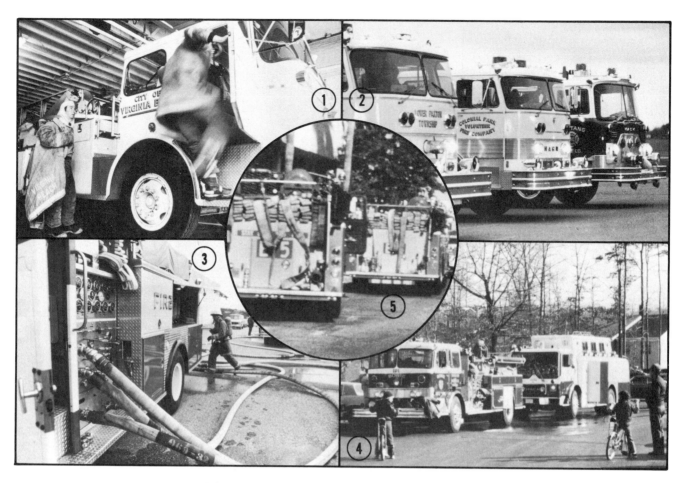

FIGURE 9.3: The five basic categories of apparatus on the fireground.

Specific assignments are directly related to these categories of apparatus status on the fireground.

Responding

All firefighting operations begin with apparatus response. Response must be conducted in a manner that allows for personnel and apparatus to arrive quickly and safely at the fire scene. Leaving quarters with crew members running after the rig and then flapping in the breeze is an unnerving and highly hazardous beginning. Firefighters should *never* run for a moving rig. Company Officers should be certain that everyone is on board or strapped in before leaving.

Officers and drivers should know where they are going, how to get there, and what other companies are responding. Some very basic information must be considered including the day of the week, time, weather, possible areas of traffic congestion along the standard route (including bridges, tunnels, schools, shopping centers, railroads, and detours), and alternate routes to the scene.

Unfortunately, there are still members of the fire service who consider a quick response to be more important than a safe one. Emergency responses have one basic and very important objective—to move personnel and apparatus safely to the incident scene. Breakneck driving is a thrilling experience for some, but it often fails to deliver the unit to the scene successfully. Those driving 70 mph usually ar-

FIGURE 9.4: A Jesse James get-a-way!

rive only seconds before their more sensible counterparts but they need Valium® to settle down after the wild ride. The FGC must do whatever is necessary to control these thrill-show drivers and to ensure that everyone leaving the station arrives at the scene safely and calm enough to begin work.

It is often a waste of time to give complex assignments to responding units by radio before they arrive on the scene. Such activities may increase response urgency and lead to unsafe driving and add confusion on the part of the personnel en route. Responding companies cannot go to work until they arrive. The FGC should wait until the company is close to the scene before giving orders, if necessary.

Staging

In order to properly maintain fireground law and order, it is essential to manage uncommitted apparatus. *Staging provides an orderly link between response and initial fireground operations and is the basis for unit deployment.*

The initial arriving attack team (e.g., an engine, a ladder, or a rescue squad) should go directly to the scene, take standard positions, assume command and begin operations. Units arriving after the initial attack team should stop about one block from the scene, remaining uncommitted until ordered into action by the FGC. Their positions should allow a maximum number of tactical options, yet not impede access to the scene. This standard procedure is called *Level I Staging.* Companies arriving in Level I staging should advise the FGC of their arrival and position by radio (e.g., "E-1 staged west").

When the FGC is faced with large, complex, or lengthy operations, he should consider staging additional units together in a specific location under the command of a Staging Officer *(Level II Staging).* The Staging Officer serves as a communications link to Command and coordinates specific assignments for the staged companies. Level II staging is most typically used during large fires where a tactical reserve is required close to the scene. This procedure is particularly effective in managing multiple alarm companies. Companies arriving in the Level II staging report to the Staging Sector Officer who is in communication with Command.

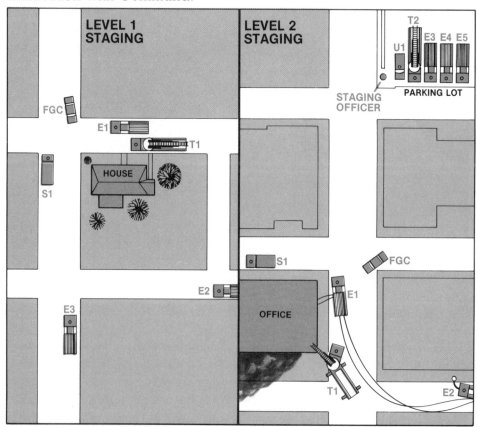

FIGURE 9.5: Level I staging is appropriate for small fires. Large, complex, or lengthy operations require Level II staging.

Once the incoming units are staged, the FGC should coordinate specific assignments to companies as they are needed with the staging officer. This is the basis of the entire staging process. As needed companies receive their orders, they are given a location and objective that integrates them into the overall strategy. This process places companies in the most advantageous spots and helps to eliminate premature commitment (free enterprise).

Operating

Working units should be placed in a manner that maximizes their effectiveness based on the critical factors of access, building characteristics, fire progress, exposures, and the location of other companies. Different types of apparatus require different placement.

REMEMBER: Basic placement decisions depend on the type of apparatus and its fireground assignment.

Parked

Many fireground assignments are personnel intensive, i.e., they do not require the direct use of apparatus. Instead, the rig has served the very legitimate function of a taxi to deliver personnel to the scene.

Apparatus should be parked to prevent congestion, yet where it can be moved into operating position if needed. Company officers must take the steps necessary to avoid being hemmed in by hose lines, other apparatus, parked vehicles, or equipment. Walking the last half block often eliminates congestion and allows the rig maximum tactical options.

NOTES

FIGURE 9.6: Park apparatus to avoid congestion, ready to be moved into operating position.

Returning To Quarters

Once companies have finished their assignments and are no longer needed by the FGC, they can be placed "available" and returned to quarters. For routine operations, this should be a simple and rapid process. However, during major fireground operations, many units will be stripped of equipment and tools, trapped by hose, and indefinitely committed to key positions (like the end domino). Releasing units from these operations should be managed by the FGC and Sector Officers. Since it is essential for companies to be placed back into service in an efficient manner, the FGC must see that all units are placed "available" as soon as possible.

KEY POSITIONS

Effective apparatus placement must begin with the arrival of the first units. The placement of the first engine, ladder, and rescue companies

207

(initial attack team) should be based upon size-up and general conditions. These first-arriving companies should place themselves quickly and to maximum advantage. Often key operating positions are available only once. If they are not filled initially, they are generally gone for the duration, placing everyone at a disadvantage. The management of these key positions is a major element in effective apparatus placement.

Key positions are those spots on the fireground that place apparatus in the best position to work to capacity. These positions offer the maximum operational advantage to the FGC. Such locations put apparatus where they can reach whatever is critical to the operation of each unit. (hose lines can reach the fire, ladders can reach access points, etc.) and allow the FGC to expand operations with later-arriving companies.

REMEMBER: Key positions allow the FGC to take full advantage of apparatus.

Key positions generally utilize either forward locations directly involved in fire operations or support positions further back. In these support positions, later-arriving units build on the initial plan and reinforce the forward units.

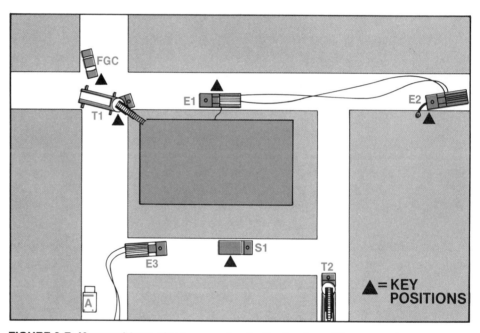

FIGURE 9.7: Key positions place apparatus in the best position to work to capacity.

The management of key positions for each basic type of unit requires different considerations and decisions.

Engine Companies

Relating key positions to the initial spotting (placement) of pumpers is an essential element in fireground operations. The Engine Officer should place his unit in a forward attack position after considering how his attack crews will enter the building. Typically, his concern will be with entry through the building's natural openings (doors and windows). At the same time, he must determine effective hose line management based on fire conditions.

In offensive situations, forward engines should be placed so that crews can quickly advance preconnected lines into the building. The location and extent of the fire must be considered to allow for protection of the unburned portion of the building, thereby assisting the primary search and control of interior access. During defensive operations, the forward engine locations should be safe ones that provide for exposure protection and the use of master streams.

REMEMBER: Spotting initial pumpers establishes the basis of the overall fire attack plan.

Key forward pumper positions must be adequately reinforced by other pumpers placed at key water sources (hydrants, drafting ponds, etc.) connected by sufficient supply lines. The standard "wagon-pumper" configuration provides a strong hydraulic reserve and allows Command to assign later-arriving companies to utilize additional attack lines from the forward pumper(s). This avoids bringing another pumper to the scene, maximizing use of the water supply and reducing congestion. The worst alternative is, a "daisy chain" of supply lines to several pumpers, which would soon outrun its own capacity.

The operation of forward key engine company positions includes the management of hydrants. As mentioned earlier, a strong fire attack requires a strong water supply. This should include placing a pumper at each hydrant to pump supply lines to the forward positions. Large diameter hose and/or multiple supply lines add efficiency to

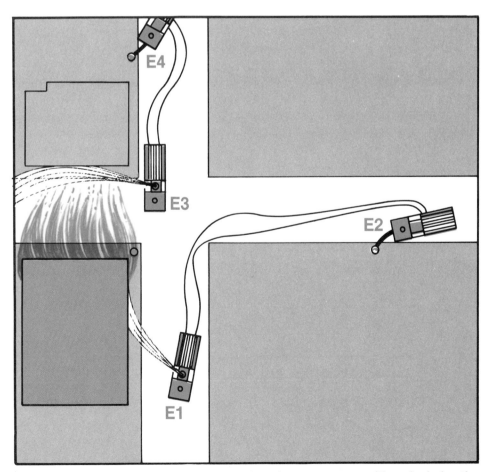

FIGURE 9.8: The operation of key forward engine company positions includes the management of hydrants.

pumping relays. Proper use of key hydrants also reduces the number of lines which must be laid, thereby reducing fireground spaghetti.

A four-way hydrant valve permits the first-arriving engine to lay a supply line and begin the attack while a later-arriving pumper takes the hydrant and increases the water supply without interrupting the flow. The forward engine can then distribute water to hand lines or master streams. Obviously, water supply information available from a preplan will have a major impact on command operations.

Key hydrants are those located closest to the fire, in safe positions, and capable of supplying reasonable flow volume. Hydrants close enough to the building to possibly jeopardize apparatus or firefighters because of structural failure or fire conditions should be avoided. Larger diameter hydrant connections on large capacity pumpers further aid in supplying multiple/large diameter lines. Laying single, unpumped lines from these hydrants reduces or destroys this capacity. Effective apparatus placement takes full advantage of key forward positions and provides additional attack lines from forward engines that already have a good water supply.

REMEMBER: Do NOT hook up to hydrants that are close enough to endanger personnel or apparatus. Do NOT use small, single, unpumped hydrant lines.

Ladder Companies

Key ladder company positions should place aerial apparatus in a proper location to assist rescue and fire control. These positions should take advantage of the best elevated access based on the type of aerial device involved. Ladder companies that provide elevated access in key forward positions can serve as the focal point for a number of companies operating in that critical tactical area. The FGC must coordinate the effective placement of aerial apparatus, remembering to consider the safe reach of the ladder or boom.

DON SELLERS

FIGURE 9.9: You can stretch another length of hose, but you can't stretch a ladder.

Since aerials are generally the largest and most stationary rigs on the scene, the management of ladder location and operation requires careful planning. Keep in mind that this may require saving the good ladder spots for later-arriving companies. Poor initial apparatus placement may prevent efficient ladder placement later. It is very difficult to move a pumper when it is flowing water.

The use of elevated streams and ladder pipes requires supply lines. The FGC must coordinate this water supply function. When hose lines are attached, the ladder becomes more stationary; another reason for correct initial spotting.

Raising of the aerial is always done for a specific reason; if you don't need the "stick," leave it in it's bed. When ladders are not needed for upper level access or rescue, they should be placed in defensive positions that provide a base for ladder pipe operations should the suppression effort turn defensive. Such a location protects the vehicle by placing it in a less exposed position, yet allows for the rapid use of ground ladders and other ladder equipment. In spotting apparatus, Ladder Officers must consider the extent and location of the fire, the most dangerous direction of spread, confinement, exposures, overhead obstructions, and structural conditions.

FIGURE 9.10: Efficient ladder company apparatus placement.

Rescue Companies

Rescue crews should position themselves in key spots that allow for the most effective rescue and treatment of fire victims and firefighters. Often these positions can be farther back than forward attack pieces but still in a location allowing direct access to and from the immediate fire area. This "parking space" should not block apparatus movement or interfere with firefighting operations.

Effective medical care cannot be rendered in an unsafe area. Environmental conditions can also limit emergency care. For example,

exposure to high temperatures will interfere with efforts to help a heat stroke victim, and there is little point in caring for smoke inhalation victims in the smoke.

It is usually more efficient to establish a central treatment (triage) area for all victims rather than try to manage patients all over the fireground. Parking lots, yards, and wide streets provide good triage-treatment points.

The rescue vehicle or ambulance, with its supplies and communications capabilities, becomes the focal point for emergency care and gives the FGC a single unit to act as a Treatment Sector.

REMEMBER: Ambulances must have access to the treatment area to facilitate care and transport. Locations on the edge of the immediate fireground area, where ambulances can move in and out without driving in front of the fire, offer improved access.

Command Vehicle

The command vehicle is the basic management location for the FGC. This is the place on the fireground where he opens an office. Generally, the better the location, the better the operation. The command vehicle, whether a car, van, bus, or fire truck, becomes the command post. From this position, the FGC should have maximum visibility of the fire building and surrounding area. The location should be a logical one that is easy to find. Ideally, this spot should be in a conspicuous position in front of the building with a view of two sides (a quarter shot). Stationed at this command post, the FGC must be able to observe the general effects of the companies working on the fire.

For normal fire operations, driveways, parking lots, and yards directly across from the fire building are good command spots. The FGC is now separated from the fire by at least the width of a street, where he is usually safe and does not block apparatus movement.

There is a dangerous temptation to focus on the obvious, the fire, and to lose sight of the overall operation—particularly when the command vehicle is too close to the blaze. Effective fire evaluation balances what is burning with what is left to burn. If tunnel vision overtakes Command, his overall evaluation abilities will be impaired. In other words, the bigger the fire and the more companies involved, the farther back the Command Post should be. The FGC must maintain a "panoramic" view of the fireground.

When setting up a command post:

* Provide for maximum visibility of the fire building (attempt to have a clear view of two sides)
* Provide for maximum visibility of the surrounding area
* Be in a conspicuous position, easy for fire personnel to find
* Be in a safe position
* Avoid blocking apparatus movement and fire operations
* Be able to observe the general effects of companies operating on the fire.

REMEMBER: The position of the command post is critical.

ATTACK TEAMS

Correct apparatus placement supports the formation of groups of operating units, or attack teams, or task forces who work together in standard positions or at tasks too large for single companies. Attack teams generally involve two engines, a ladder or rescue squad, and a command officer. Sometimes the team will arrive and immediately begin working together. In other scenarios, Command may start with a single company and build the team with later-arriving personnel, particularly during rural operations. This team approach provides the basis for the early decentralization of fireground command. It is the beginning of the fire management organization. Starting from his arrival, the FGC should identify tactical requirements and position apparatus in a manner that builds both teams and sectors.

STANDARD STAGES OF OPERATIONS

Assigning and placing responding units involves balancing the number of companies on hand with the amount of work to be done.

There are four standard stages that will influence how companies are assigned and placed. They are the:

1. Deficient resource stage
2. Adequate resource stage
3. Standby stage
4. Fatigue stage.

DEFICIENT RESOURCE STAGE

During the initial stages of active fires, there are generally more jobs to do than companies to do them. Companies are assigned tasks as soon as they arrive. These rapid assignments during the early stages usually will be first to key forward positions and then to back-up functions. Excessive congestion is not usually a problem at this point, unless someone has "screwed up" a key position. This is an exciting stage, with everyone moving quickly. The challenge is to get the proper units into the critical forward spots.

ADEQUATE RESOURCE STAGE

When sufficient units have arrived to cover both the key forward and back-up jobs, the adequate resource stage starts. As the FGC begins to cover the entire fireground, the assignment process slows down and the operation assumes a full-employment equilibrium between tasks and companies. Good placement will now allow a rapid, flexible response to changing conditions and permit maximum use of all personnel. By contrast, poor placement will result in congestion and reduce the effectiveness of the collective effort.

STANDBY STAGE

At this point, the working companies are operating in their assigned places and are in good shape, making adequate progress. All assignments are basically stable. If the FGC has more companies than jobs, these uncommitted units can be held in staged positions as a tactical reserve. The FGC must regulate the placement of these companies and personnel on the fireground.

FATIGUE STAGE

When workers begin to tire, the FGC must start to rotate units to protect the safety and welfare of the fatigued firefighters and to utilize the energy reserve of staged companies. Although the speed and urgency of assignments are reduced, good placement puts Command in a strong position to begin rotation.

These four stages describe the typical progression of fireground deployment as it affects apparatus placement. Actual fire conditions will dictate how and when this progression will occur. Some fires will not last long enough to move through every stage; some will.

REMEMBER: Placement is the consistent element in every fire response. The FGC should match apparatus placement with apparatus function and needs.

Offensive and Defensive Positioning

The basic strategy (offensive/defensive) is a major factor in vehicle placement. Offensive spots need to be close enough to the fire to provide for operations into and onto the fire building. However, the FGC also must consider the possibility of conditions changing while spotting apparatus. The positions should be as safe as possible, but they must also be within effective reach of the fire if an offensive attack is to begin.

During defensive functions, the FGC must position units in safer locations, farther from the fire, concentrating on cutoff points that will protect the exposures. Fires that start offensively, then expand to a defensive status, may require that offensively spotted vehicles be protected and moved to more secure places. Apparatus should always be spotted based on a pessimistic forecast of extension. It is dysfunctional but sometimes necessary to move apparatus during a fire.

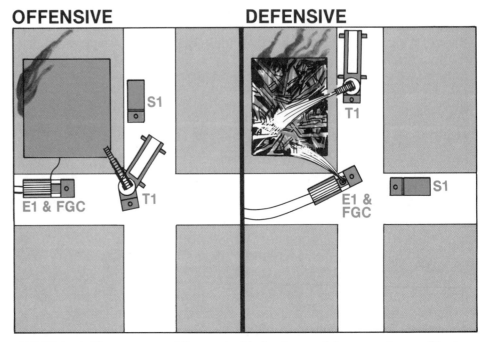

OFFENSIVE **DEFENSIVE**

FIGURE 9.11: The strategy of fire control helps to regulate apparatus positioning.

At times, it will be necessary to protect endangered fire trucks with fire streams until they can be safely moved. When conditions do change and fire begins beating up the apparatus, do not engage in long decision-making strategy sessions. Quickly place water streams on the rig and move it to a safe location.

FIGURE 9.12: Get on ol' Paint and ride where you ain't.

The cost of fire apparatus is close to the national debt and delivery time is equal to the gestation period of an elephant (two years). Based on these realities, Command should think of fire apparatus as expensive exposures. Many times the cost of the fire vehicles on the scene exceeds the cost of what is burning (sometimes it exceeds the value of everything within two blocks). It is bad form to burn up or bury a truck that was in the wrong place when fire conditions changed.

When positioning working units, the FGC must consider the extent and location of the fire and then make a pessimistic evaluation of fire spread and building failure. Also, he must anticipate the heat that may be released when buildings open up during structural collapse. Even in "nothing showing" situations, it is necessary to park units at least 30 to 40 feet from the fire building (100 to 150 feet for high-rise buildings). Many times, greater distances are required.

An ongoing evaluation of what's burning vs. what's left to burn offers a framework to predict the future of certain spots. The FGC must beware of positioning trucks where they cannot be repositioned easily and quickly or where there is only one route of access in and out (e.g., alleys, driveways, yards, narrow streets, etc.). If the FGC does make the commitment, he has to make sure that everything needed to supply and protect the unit is provided.

Another spotting hazard is overhead power lines. Everyone working should develop the instinctive habit of looking up to find power lines and their relation to the fire. Units should not park under lines if at all possible.

As the entire operation gets older, fire hose (particularly large diameter) soon limits general access. Command and Sector Officers should have everyone well placed and supplied before this happens. Lines should be laid with attention to the access problems they will present. Engine companies should attempt to lay lines on the same side of the street as the hydrant and then cross over, when necessary, near the fire. If possible, they should maintain an access lane in the center of the street to provide space for apparatus that might need to get through later.

All placement is regulated to a great extent by the general conditions of each department's response area. What works in Phoenix may never work in East Bigosh. Street width is an important example. Many western cities have 100 foot-wide streets, while most streets in Boston are typically no more than 30 feet wide, with cars parked on both sides.

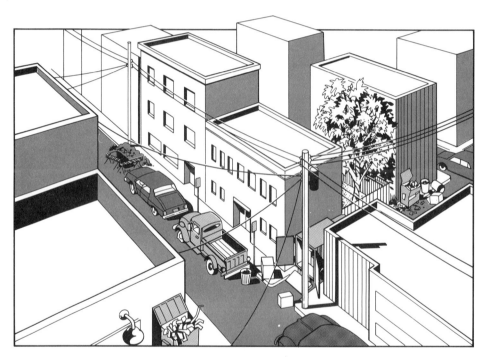

FIGURE 9.13: Effective placement of apparatus may be limited by many factors.

While positioning is not difficult and mistakes are easy to correct in a 10-acre field, narrow streets and alleys require clever maneuvering, careful positioning, and a strong fire plan. Each situation necessitates a special approach. This is where the initiative, training, and experience of the officer and driver have an incredible impact on the eventual outcome.

SUMMARY

The major goal of apparatus placement is to properly position units on the fireground in order to utilize their functions to the best advantage. Apparatus function should regulate its placement, expanding the capabilities of all companies at the scene.

Efficient apparatus placement requires a SOP for the first-arriving companies, a prearranged staging procedure, careful decisions made by the Company Officer that are based on existing or predictable conditions, and direct orders from the FGC. Tactical options will disappear if fireground congestion occurs.

There are five basic categories for apparatus on the fireground: responding, staged, operating, parked, and returning to quarters.

During response, safety is the primary concern. All personnel must arrive at the scene safely and calmly enough to receive orders and start to work.

The effective management of uncommitted apparatus is essential to maintain fireground order. It forms the basis for fireground deployment. The initial attack team should go directly to the scene, while later-arriving units should stop about one block short of the scene and wait in an uncommitted position. They must not restrict access to the scene. A central staging area with a Staging Officer is very useful for large fires and in managing multiple alarm companies.

The basis of the entire staging process is to assist the FGC in making specific assignments that include a goal and an objective.

Units working on the fireground should be placed to maximize their effectiveness. Placement decisions depend on the specific type of apparatus and its tactical assignment.

When apparatus serves as a taxi to the scene, it should be parked to prevent congestion, yet where it can be moved to an operating position (avoid being hemmed in by hose lines).

Companies should be placed back into service in an effective manner. Releasing companies should be coordinated by the FGC and Sector Officers.

Key fireground operating positions may be gone quickly if they are not assumed upon arrival. Key positions are those spots on the fireground that place apparatus in the most effective position. They generally involve forward spots close to the fire and support positions farther back.

The management of key positions require the FGC to consider effective placement of engine companies, ladder companies, rescue companies, and the command vehicle.

When considering engine companies, the FGC should place engines in key forward fire attack positions.

In offensive situations, forward engines must be placed so that crews can advance preconnected attack lines quickly into buildings. Officers must consider the location and extent of the fire and then spot apparatus where fire attack efforts will protect unburned portions of the buildings.

In defensive operations, forward engines must be spotted in safe locations that provide for exposure protection and master stream operation.

Key forward positions require adequate reinforcement by other engine pumpers placed at key water sources. A standard "wagon-pumper" configuration will provide a strong hydraulic reserve. This approach allows the FGC to assign later-arriving companies to take additional attack lines from forward engines, eliminating the need to bring another engine to the scene.

Key engine company positions include the management of hydrants, placing an engine at each hydrant to pump supply lines to forward positions. Large diameter hose and multiple supply lines add efficiency. Avoid laying single, unpumped lines (other than large diameter hoses) from any hydrant.

Key hydrants are those located closest to the fire in safe positions.

Key ladder company positions should place aerial apparatus to assist rescue and fire control. When possible, the FGC should coordinate this placement based on the length of the ladder, boom, articulating arm, etc.

Since aerials are large and essentially stationary rigs, placement requires careful planning. It may be necessary to save the best ladder spots until their arrival. Ladder companies may provide good elevated access to key forward positions, thus serving as a focal operating point for many forward operations.

Keep in mind that when hose lines are attached to a ladder, the ladder becomes stationary.

Ladders not in use for upper level access or rescue should be placed in defensive positions.

Key spots for rescue vehicles are positions that provide for the efficient emergency care of victims. These vehicles should be placed in a safe area which has direct access to the fire area and allows for transport. They should not block the movement of other apparatus. For multiple victims, a central triage and treatment area is needed.

The command vehicle serves as the command post for the FGC, situated in a safe, conspicuous position that allows the FGC to have maximum visibility of the fire building and the surrounding area. Its position should not interfere with operations or apparatus movement. The bigger the fire and the more companies involved, the farther back the command post should be.

Attack teams are assisted by efficient apparatus placement. These teams work in positions or at tasks that are too large for single companies. They are the basis for early decentralization of the fireground command.

Typical fireground deployment occurs in four stages: deficient resource, adequate resource, standby, and fatigue. During the deficient resource stage, there are more jobs than companies to do them. The adequate resource stage has sufficient units to cover key positions and back-up jobs. In the standby stage, assignments are basically stable. Uncommitted units are held in staged positions as a tactical reserve. Workers are tired during the fatigue stage, requiring the FGC to rotate personnel.

Vehicle placement is dependent upon the basic strategy of fire control. Offensive positions need to be close enough to the fire to allow extension of operations into and onto the fire building. During defensive actions, units are placed in safe positions and concentrate on cutoff points. If a fire expands to a defensive status, vehicles in forward positions may have to be protected and then moved.

When positioning units, consider the location and extent of the fire and then make a pessimistic evaluation of fire spread and building failure. In "nothing showing" scenarios, park units at least 30 to 40 feet from the fire building. Avoid placing apparatus where it cannot be repositioned quickly. Make certain that overhead power lines will not cause a hazard.

COMMAND DEVELOPMENT

In order to manage efficient and effective fireground activities, fire companies, and sectors, the FGC must be highly skilled in apparatus placement. Since there is a strong relationship between apparatus function and placement, the FGC must be familiar with the apparatus he will be commanding at the fire scene. He must know its primary functions and its limitations.

The FGC should know the basic objectives of apparatus placement and study how the apparatus he will use functions at the various scenes common to his locality.

The following report card is provided so that you can evaluate your FGC knowledge and skills in classroom exercises, simulations, and on the fire scene.

Fireground Commander Report Card

Subject: Apparatus Placement

Did the Fireground Commander:

- ☐ Regard and manage apparatus in the five basic categories?
    - ☐ responding
    - ☐ staged
    - ☐ operating
    - ☐ parked
    - ☐ returning to quarters
- ☐ Make specific assignment to companies?
- ☐ Utilize Level I staging for routine operations and Level II staging for Big Deals?
- ☐ Consider key positions for:
    - ☐ engine companies (both attack and supply)?
    - ☐ ladder companies?
    - ☐ rescue companies?
    - ☐ command post?
- ☐ Form attack teams?
- ☐ Provide adequate and reliable water supply?
- ☐ Consider offensive/defensive strategies in placement?
- ☐ Place rigs in safe positions?
- ☐ Protect apparatus as conditions changed?
- ☐ Maintain access to the scene?

10

SAFETY

MAJOR GOAL

TO DEFINE THE RULES, PROCEDURES, AND MAJOR FACTORS REQUIRED FOR FIREFIGHTER SAFETY AND WELFARE.

OBJECTIVES By the end of this chapter, you should be able to:

1. Explain how SOPs can help the FGC make safety a matter of routine. (p. 222 and 228)
2. List 10 safety factors. (pp.222-223)
3. State how fireground safety ranks in terms of the tactical priorities. (p. 223)
4. State how the FGC can help set the right attitude for safety. (p. 223)
5. List four ways the FGC can carry out risk management. (p. 223)
6. State the safety advantages of assigning sectors and Sector Officers. (p. 224)
7. Define the role of a Safety Officer. (pp. 225-226)
8. Explain how deciding the mode of operations is related to safety. (p. 227)
9. Define the "20 minute rule," indicate how it applies to offensive operations, and list its limitations. (p. 227)
10. List seven signs of structural weakness or impending failure. (p. 228)
11. List at least six tactical positions or actions that require special caution. (p. 228)
12. State two ways in which SOPs affect safety. (p. 228)
13. List at least three firefighting action areas that must be covered by SOPs as a matter of policy. (pp. 229-230)
14. Describe the general procedures for an emergency evacuation, including communications. (p. 231)
15. Define "fireground perimeter" and state who should be on the fireground at any given time. (p. 232)
16. List five factors to consider when setting up a fireground perimeter. (p. 233)
17. List four human factors that should be considered at the rehabilitation sector. (p. 235)
18. Define "full protective clothing." (p. 236)
19. List at least six situations where self-contained breathing apparatus (SCBA) must be worn. (p. 237)
20. List at least five basic rules for vehicle response safety. (p. 238)

SAFETY

THE NEED FOR SAFETY

Firefighting is the most physically demanding and dangerous profession in the United States. Fireground operations involve many inherent dangers and very real risks to the participants. These dangers include fire, smoke, toxic products of combustion, electrocution, structural collapse, explosion, vehicle accident, stress, overexertion, equipment failure, and the direct results of uncoordinated tactical activities. Fireground action should be the domain of those participants who are physically fit, mentally alert, healthy, properly trained, fully protected and equipped, and organized to perform in a safe, coordinated manner.

The FGC must always remember that *the safety of the participants is a primary goal in all phases of every tactical situation*. The risks that come with the fire are the reason that the citizens call the fire department when they need help. The FGC must recognize, manage, and reduce these risks. On the other hand, the risks created by unsafe actions, attitudes, and performance of the fire suppression troops are intolerable and must be eliminated through the FGC's constant and conscientious management of fireground safety.

The FGC is directly responsible for the safety of each firefighter as well as for the overall safety of the entire operation. This concern for safety must saturate every level of the organization. A healthy approach to safety begins to take effect when the individual participants recognize their responsibilities for their own safety and for the safety of their coworkers. Nothing makes this happen more effectively than the FGC who displays a firm and healthy attitude toward safety at all times. When the FGC takes safety seriously, everyone else does too.

THE SAFETY PROGRAM

A proper approach to fireground safety must be based on a program which includes SOPs directed specifically toward safety practices and risk management. When the SOPs are structured around operating in the safest manner, safety becomes routine and everyone learns to expect it. The regular game plan provided by SOPs defines standard activities and expectations. The most unsafe operations are those where no one has a plan or a standard approach to the problem. In such cases, the out-of-control mode overcomes rational activities, with chaos defeating safety, every time. The FGC must never tolerate a reckless approach to fireground operations.

Safety Factors

The critical elements of a fireground safety program must include:

- **COMMAND ATTITUDE**—Safety is a primary responsibility of the FGC and the command organization.
- **FIREFIGHTER ATTITUDE**—Everyone involved in the operation is concerned with safety and accepts personal responsibility.
- **STANDARD OPERATING PROCEDURES**—Safety is built into every operation as a standard approach.

- **SECTOR OFFICERS**—All fireground operations are directed and coordinated by Sector Officers with a standard responsibility for managing safety.
- **SAFETY TRAINING**—All personnel are trained in safety practices, procedures, and approaches.
- **HEALTH AND FITNESS**—The participants are in the proper physical and emotional condition to perform as required without compromising their own safety or the safety of others.
- **SELF-CONTAINED BREATHING APPARATUS**—Nobody operates in a contaminated atmosphere, or one which may rapidly become contaminated, without full respiratory protection.
- **PROTECTIVE CLOTHING**—Everyone wears full protective clothing, whenever subject to physical hazards.
- **EQUIPMENT AND APPARATUS**—Well-maintained, properly designed, and up-to-date apparatus and equipment are provided to do the job safely and effectively.
- **RISK MANAGEMENT**—A limited amount of risk is accepted as part of the job, but no effort is spared to avoid or control unmanaged risks.

Attitude

The FGC is the one person who has the overall responsibility for fireground safety. *The fourth basic fireground priority is the safety of the participants, ranking equally with rescue, fire control, and property conservation.* Unlike the others, safety does not have a time when it becomes important in the order of fireground activities. Safety must always be important throughout the entire fireground operation.

If the FGC tolerates a relaxed attitude toward safety during day-to-day operations, he will never be able to gear up the safety system when something particularly dangerous comes along. When the participants begin to think that safety is not important during routine incidents, they are subjecting themselves to unknown and unanticipated disasters. Many fireground injuries and deaths occur during these times, because standard procedures and precautions are simply ignored. An incident that is "no big deal" can turn very bad when the slightest little routine detail suddenly changes and nobody is prepared for it.

By the same token, some FGCs throw away the safety plan (and every other component of rationality) when the big one comes along. "This is no time to worry about safety" is their battle cry. These officers generally have extensive experience in arranging big funerals.

Command Officers must demand that the professionalism of fireground participants includes a healthy respect and concern for safety and display these attitudes by setting a positive example.

Risk Management

In spite of every appropriate precaution and procedure, the fireground continues to be a risky place to do business. The FGC can recognize and anticipate danger, organize the fireground, assure that safety procedures are followed, and demand a safety conscious attitude from his personnel, but he can never eliminate the component

of risk from fireground operations. The important question is how much risk the FGC is willing to tolerate.

The *hero mentality* is the most serious mental illness in the fire service. Firefighters become heroes by risking their lives to save the lives of helpless victims. Sometimes those efforts are successful and sometimes they are not . . . sometimes the rescuers become victims themselves. It makes very little sense to risk the lives of firefighters trying to save victims who are already dead. It makes even less sense to risk firefighters trying to save unoccupied property, and it is absolutely insane to expose them to danger in an abandoned building.

A realistic approach to risk management should include the following axioms:

- Savable victims are the absolute number one priority of all fireground operations.
- Victims who are already dead are not savable.
- Victims inside fully involved fire areas are generally dead within less than one minute.
- No property is worth the life of a firefighter.
- Contents that are already on fire have very little salvage value.
- If we save an abandoned building today, someone will burn it tomorrow.
- A lot of the stuff we risked our lives for on Saturday night gets loaded into an old dump truck and hauled off Monday morning.
- Beware of crews that always attack and know only one pace—full speed ahead.

How the FGC views risk is critical in the way he manages the fireground. If he has a "hero mentality," his crews are in trouble. If he doesn't let his people fight the fire, the citizens are in trouble. Managing risk does not mean that the FGC lets every building burn because he is afraid to expose personnel to any danger. It does mean that he regularly asks himself if the risk to his personnel is justified by the results they can achieve. He must be a realist, keeping in mind that the basic business of the fire department is protecting lives and property.

STANDARD SAFETY PROCEDURES

FIREGROUND ORGANIZATION

Safety is one of the primary reasons for organizing the fireground and assigning responsibility to Sector Officers. Sector Officers are directly responsible for the safety of personnel and operations within their assigned areas and for coordinating activities with other sectors and the FGC to prevent confusion. This level of coordination avoids situations where the attack crews from the Rear Sector blast the fire toward the crews coming through the front door and keeps the Roof Sector from opening-up before attack lines are in position.

Each Sector Officer must be aware of the position and function of every crew assigned to the sector and must be confident that they are operating safely and using proper protective equipment. This requires the Sector Officer to be dressed and protected appropriately so that he can be inside where the companies are operating, seeing what they are seeing, and knowing what they are doing. If the companies need to be in full protective clothing and SCBA, the Sector Officer needs to be setting the example.

This level of organization becomes very important when things begin to change rapidly on the fireground. If the strategy changes from offensive to defensive, or if something suddenly goes wrong, the FGC needs to immediately change the position and function of companies and rapidly account for their welfare. Sector Officers must be in position to implement immediate changes and account for their personnel, just as each Company Officer must be constantly aware of the position and function of his crew members. In situations which present an unusual degree of risk, Sector Officers should account for each individual in the danger area through the use of a *personnel identification system*. This allows them to know, at a glance, who is in and who is out, by collecting and returning name tags carried by each participant.

NOTE: The British fire service routinely employs a system to postively account for every individual on the fireground using breathing apparatus. Each SCBA comes with a tag indicating the identity of the user and the air pressure at entry. As the firefighter enters the work area, the tag is given to the Breathing Apparatus Control Officer who monitors the times of entry and expected exit. This provides full accountability for everyone inside the fire area.

When sudden changes occur on the fireground, the FGC should immediately check with each Sector Officer. The Sector Officers should be able to give an immediate report on the welfare of all assigned crews. During more routine times, the FGC should be in a position to observe any unsafe conditions or actions and react immediately to them. Decentralizing command responsibility to Sector Officers places the FGC's eyes and ears where the action is, looking out for the overall safety concerns.

The Sector Officers also look out for the welfare of personnel by monitoring their fatigue levels and making sure that they are rotated, replaced, or sent to a rehabilitation area before they reach the point of exhaustion. Stress and overexertion can lead quickly to the point where firefighters become dangerous to themselves and to their coworkers.

THE SAFETY OFFICER

The Safety Officer is a specialist who provides a higher level of expertise and undivided attention to supplement the FGC's role and responsibility for fireground safety. In larger fire departments, the role of Safety Officer may include one or more full-time staff officers committed to managing the safety program, in addition to responding to emergency scenes. Where there is no full-time Safety Officer responding, every officer should be prepared to function as the incident Safety Officer when assigned by the FGC. The presence of a Safety Officer does not replace common sense and in no way diminishes the responsibility of individual fireground officers for safety—it adds a higher level of attention and expertise to help them.

The Safety Officer is essentially an *advisor* to the FGC and a *consultant* for the Sector Officers. A full-time Safety Officer brings to the fire scene an officer who has particular expertise in analyzing safety hazards and investigating accidents and injuries. The Safety Officer also should be the person most familiar with the uses and limitations of protection equipment and the requirements of safety procedures.

While the Safety Officer normally functions as an advisor to the FGC and represents safety through the normal chain of command, he also has the overriding authority to "blow the stop whistle" or veto a plan when conditions or actions create an immediate safety hazard. A Safety Officer must resist the temptation to use this authority "Chicken Little" style, but everyone on the fireground, including the FGC, must realize that when the Safety Officer says "evacuate" it is not the time to compare seniority dates or call for a vote.

The Safety Officer should function consistently as a *sector* within the fireground organization. This establishes safety as an equal element in the command structure, reporting directly to the FGC. This is a functional sector assignment with full authority to move around the fireground to represent safety concerns. One of the principle strengths of this approach is the ability to stand back and look at the big picture. The Safety Officer can look at the overall situation and action with a certain degree of detachment, since he is not preoccupied with trying to achieve a particular tactical objective. It is often amazing how much can be seen by just watching what is going on and looking at the fire and structural conditions.

The size and complexity of an incident may require more than one person to satisfy the needs for Safety Sector attention. Additional officers may be assigned to the Safety Sector to monitor the overall scene or to provide a direct level of supervision over a particularly dangerous part of the action. One or more companies also may be assigned to work under the Safety Sector to monitor a specific hazard or to enforce specific requirements.

Specific expertise can be added to the Safety Sector's capabilities by assigning a fire protection engineer, structural engineer, or hazardous materials expert to this area of responsibility. When personnel are involved in highly technical situations and when special hazards or dangerous structural conditions are present, the FGC should be prepared to call upon the best help available.

The Safety Officer has a responsibility to represent safety policies, procedures, and requirements on the fireground. Unfortunately, this often requires corrective or regulatory actions which create a "safety cop" image. The role of the Safety Officer is meant to be for the welfare of personnel operating under the stress, excitement, and danger of the fireground. This role must not be compromised by irrational attitudes.

The FGC must recognize incidents and activities which create an unusual degree of risk to personnel and then react accordingly by increasing the level of safety supervision. *A basic level of safety is always in effect, even in the most routine situation.* As the intensity of activity advances to a working fire stage, the FGC reacts by assigning a Sector Officer to manage safety as a singular concern. A third and higher level of safety supervision is implemented when personnel are engaged in particularly hazardous activities. In these cases, the direct safety supervision is further increased with the Safety Officer and additional supervision applied directly to the hazardous action. The position and specific function of each individual is subject to direct control using this system.

STRATEGIC AND TACTICAL CONSIDERATIONS

The safety of firefighters is one of the critical factors in deciding if a fireground operation is to be in the offensive or defensive mode. If the fire obviously outmatches the fire control capability of an interior attack, the decision to go defensive is clear. In many cases, however, the decision must be based on safety as well as capability. The FGC may decide to implement a marginal strategy—send crews in to attempt an interior attack but be ready to pull them out and go defensive if things start to look bad. In these cases, there must be no hesitation to pull out the crews and go with the big guns if safety is compromised. To make the right decisions, the FGC must know where his crews are operating and what they are doing.

An aggressive interior attack can effectively eliminate many safety concerns by extinguishing the fire, but if the fire is not quickly controlled, those safety concerns will rapidly become critical. If the fire is winning and the troops are losing, the FGC needs to get them out. When they do pull out, the fight becomes defensive, operating heavy streams from safe positions to protect exposures, not operating hand lines in doorways to protect the wooden door frames. When everyone is out and accounted for, the master streams can open up to blast the fire. There must be no mixing of modes; interior crews never react kindly to being deluged by a two-inch straight tip or the sudden reversal of flowing heat and smoke.

> **REMEMBER: If in doubt, the FGC must make the decision to go defensive. There may be an opportunity to reverse the decision after further evaluation and go back in, but there may be no second chance for firefighters who stay in too long.**

One of the real concerns for the FGC is how long the structure can be expected to hold together under fire conditions. A fire resistive or heavy timber building may stand up to prolonged assaults, but other types will begin to come apart much more rapidly.

The old *20 minute rule* told us that we could expect most ordinary construction buildings to withstand about 20 minutes of heavy fire involvement before failure. Based on this assumption, the FGC would be able to set a timer for 15 minutes, make an evaluation when the bell rang, and still have 5 minutes of safety margin. Unfortunately, buildings do not come with any 20 minute guarantee certificates, so the FGC must constantly monitor and re-evaluate the safety of the structure and use the rule as a way to maintain an awareness of time in relation to structural conditions. Many newer types of lightweight construction will not last even 10 minutes, and all types of buildings may contain hidden flaws.

In addition to everything else, the FGC needs to maintain an awareness of time—how long the building has been on fire, how long the crews have been inside, and how long it will take to accomplish various tactical options. Time keeps marching on, so the FGC has to take advantage of a situation while it lasts. When time runs out, the FGC runs out of options.

While he is constantly re-evaluating attack options, the FGC should be looking out for the signs of structural weakness or impending failure, including:

- Leaning
- Cracking
- Twisting
- Flexing
- Groaning
- Leaking (smoke or water through walls)
- Disappearing roof-mounted equipment
- Bricks landing in the street.

Certain tactical positions and actions are particularly risky and may present an unusual danger to firefighters. When crews are operating in these positions, the FGC needs to be especially concerned with their safety and must be prepared to react to negative reports. Situations requiring special caution include:

- Crews operating directly over the fire (roof or upper floor)
- Attack positions where the fire can get behind attack crews
- Roof structures which may fail suddenly
- Below-ground fires
- Any unvented interior fire
- Where Sector Officers do not have direct control of position and function
- Limited access/exit situations—only one way out
- Incidents involving hazardous materials
- Potentially opposing attack directions
- Exterior attack combined with interior attack.

STANDARD OPERATING PROCEDURES

Standard operating procedures affect safety in two different ways. The first and most basic reality is that when everyone is operating within a format structured by SOPs, surprises are eliminated and everyone has a good idea of what should be happening, who should be doing what, and how it should be done. These basic understandings reduce confusion and increase safety by helping to keep everyone on the same game plan. Situations where nobody has a plan and everyone is in action are absolutely dangerous. The FGC needs to have a strategy and an attack plan; everybody else needs to understand what these are and how their actions relate to the plan. When plans and actions are structured around SOPs, and the SOPs contain safety considerations, the entire operation has a positive start.

The second aspect of SOPs is the contribution of procedures directed specifically toward safety. These SOPs define regular requirements which apply to the actions of firefighting personnel, over and above any other procedures. These SOPs are *absolutes*, defining the rules which must always be followed, regardless of strategic decisions and tactical options. There are no escape clauses or discretionary judgments in mandatory safety SOPs.

Some of the areas which must be covered by SOPs as a matter of policy include:

- *Protective clothing*
  - Defining what types of protection must be worn on the fireground and specifically limiting options to those which provide acceptable head-to-toe protection
  - Requiring protective clothing to be maintained in proper condition
  - Specifying additional safety equipment which must/may be carried (flashlight, pry tool, rope, etc.).

FIGURE 10.1: Full protective gear is to be worn at all times when operating on the fireground.

- *SCBA*
  - Defining where breathing apparatus must be used—any location where the atmosphere is, or may be, contaminated or oxygen deficient, and anytime that explosion or structural failure would expose the user to respiratory dangers
  - Defining where breathing apparatus must be worn and available for immediate use—any location where the atmosphere could rapidly become contaminated
  - Defining when face pieces may be removed—assurance that the atmosphere is clear
  - Specifying inspection and maintenance—daily inspection and regularly scheduled maintenance.

FIGURE 10.2: SCBA drastically improves firefighter safety and efficiency.

- *Emergency Response*
  - Rules which provide for a safe arrival at the scene, satisfying legal requirements and outlawing irrational driving
  - Driver/operator qualifications and training
  - Permissible riding positions—in seats with seat belts
  - Vehicle maintenance and inspection.

Additional safety SOPs should cover areas which would be applicable to a specific fire department's operations.

There must be a standard system established in order to identify who is on the fireground so that the FGC can account for all the players if something goes wrong. Every officer needs to know his fellow crew members, and everyone must be accounted for at the scene. This is a particular problem for some volunteer departments where the players come from all directions when the siren blows.

A fireground operation structured by standard operating procedures is necessarily a safer operation than one where the free enterprise system and "anything goes" tactics are tolerated. When crews go in to perform search and rescue, they must do so as organized teams with specific assignments. The crews who are searching must be coordinated with the guys who are advancing hose lines and the truckees who are opening the roof. If everyone swarms into the building and does whatever feels appropriate and urgent, using their own tactics, the FGC becomes a spectator to an out-of-control situation, unable to effectively influence the outcome or provide for the safety of his own firefighters. The FGC who keeps calling for help so that he can send more and more players into the battle, hoping that they will do the right combination of things to overcome or overwhelm the fire, is giving up his responsibility to bring them back alive. The structure provided by SOPs is the first step in the safety system.

Evacuation

One of the most critical and urgent situations which can occur on the fireground is when the FGC has to evacuate a fire building and switch to defensive strategy. When things are working well, this decision can be made in a timely manner and crews can be withdrawn and repositioned cautiously and deliberately. Sometimes, however, things will start to turn bad very quickly, and the FGC must make an immediate decision to pull the troops out in full retreat. In these cases, future tactical considerations become unimportant and the entire emphasis shifts to getting the attack crews out of danger. The entire approach to organizing and managing the fireground becomes the important basis for moving inside firefighters outside during situations that require quick evacuation. The regular system of routinely assigning a controllable number of firefighters to a Sector Officer gives the FGC the ability to use that Sector Officer to move those firefighters out of the building when a tactical situation quickly goes from a winner to a loser.

A distinctive *emergency traffic* tone over the radio, coupled with an "evacuate the building" message from the FGC to Sector Officers in exposed areas, pre-empts any other plan and becomes the immediate priority. There is no discussion or debate at this point; everyone goes for the exits without delay. Hose lines are either backed out (to protect crews as they exit) or abandoned on the inside and used as life lines. Company Officers make sure that their crews get out, and Sector Officers make sure that their companies get out.

Once they are out and away from the danger, Sector Officers must assemble their troops, account for everyone, and report an "all clear" back to the FGC as soon as the count is made. The only significant consideration at this point is making sure that everyone is safely out of the danger area. Knowing the position and function of everyone on the fireground before the emergency is the basis for establishing their safety when things go wrong. Under the worst circumstances, it provides the foundation for organizing a rescue effort.

Missing fire personnel on the fireground cause everyone to forget everything they have ever learned about safety and try any desperate tactic to do something (anything!). In fact, this is when organization and carefully controlled action becomes most important. A rescue effort may require the coordinated actions of everyone on the fireground under a rational plan. Irrational and uncoordinated efforts may only prove that a bad situation can be made worse. The FGC must assume an extremely strong leadership during these desperate times.

Sudden shifts from offensive to defensive strategy can be so disruptive to action that the attackers cannot be quickly regrouped into defenders. While they are picking themselves up and untangling their attack lines, the FGC may elect to bring in a whole new team (preferably from a staging area) with their own apparatus to take defensive positions.

Evacuation is a SOP which should always be available but seldom used. An FGC who frequently has to implement this option is not ef-

231

fectively evaluating conditions and forecasting outcomes. If the troops figure out that the FGC is in the habit of sending them into places where they have to run back out, they will quickly lose confidence in him. By the same token, a Command Officer who keeps finding his initial attack crews in untenable positions needs to have a serious discussion with the Company Officers.

DEFINING THE FIREGROUND

Many of the standard safety requirements are considered to always be in effect on the fireground. This requires that everyone involved understands where the boundaries of the fireground are located. For safety purposes, the fireground is defined by an imaginary line which encloses the space where the fire situation creates a potential hazard to personnel. The definition of a *standard fireground perimeter* avoids any confusion or conflict over where the specific requirements are in effect and where they are not.

Nobody should be inside this perimeter unless they:

- Have a specific assignment or function to perform
- Are wearing full protective clothing
- Have their SCBA
- Are functioning with their assigned company or crew
- Are assigned to a sector.

ALL OTHERS STAY OUTSIDE!

The rigid enforcement of these rules virtually eliminates freelancers and wanderers in the area where they are exposed to danger.

FIGURE 10.3: The fireground perimeter should be no mystery to the players.

Based on the reality of most fire situations, the fireground perimeter should be no mystery to the players. If the fire building is surrounded by wide streets and parking lots, the center of the street and an equivalent distance in the clear space around the building is generally sufficient. If apparatus is properly placed (see Chapter 9), the apparatus closest to the building can usually be used as markers—between the apparatus and the building is "IN," beyond the apparatus is "OUT." If this open space is not available, the perimeter may have to include the entire street or alley and the exposures on either side of the fire building. If apparatus is operating within this area, the operators must be properly protected.

The absolute consideration for the perimeter should be how far the potential danger extends from the fire. The factors considered when setting the perimeter must include:

- Areas subject to structural collapse
- Areas of potential explosions
- Areas of smoke drift
- Areas of falling debris (a basic perimeter 200 feet in all directions should be maintained around high-rise buildings)
- Location of the fire in relation to the center of the street and clear areas available around the building.

The majority of structure fires do not create any major challenge in identifying the fireground perimeter, even with imaginary lines. Where the hazards are not clearly evident, such as hazardous materials incidents or areas where a weakened part of the structure may be expected to fall, the FGC should have rope or banner tape stretched to define a visible boundary line. This boundary line defines a special hazard area and a perimeter which may only be entered through an access control point. (If no access control point is provided, nobody is authorized to go there.) The definition of these *special hazard zones* should be a standard function of a Safety Sector.

The command post should be outside the fireground perimeter and sufficiently removed from hazards so that the FGC can operate without his SCBA and gloves. Sector Officers operate inside the perimeter, where the action is, so they must be fully protected. Only the minimum number of personnel should be inside the perimeter and exposed to hazards—particularly in high-risk situations, and any crews which are not being used should be sent to rest or to standby outside the perimeter. Basic safety requirements are still in effect outside the perimeter, but the stringent requirements for operating within the fire area may be relaxed.

Beyond the fireground perimeter is an additional area, generally controlled by the police and defined by a *fire line.* This area is reserved for the fire department to operate without having to deal with spectators, traffic, and other problems. Within this intermediate area, personnel should be expected to operate routinely with a normal approach to safety, sharing the space with reporters, photographers, ambulance drivers, police officers, mayors, city managers, fire buffs, visiting firemen, and owners of burning buildings, all of whom may be officially authorized to be there. The limits of the fire line should be established through a standard arrangement with the police department and confirmed by the Police Liaison Sector.

While the hazard created by the fire should be limited to the fireground perimeter, personnel operating within this outer zone, particularly apparatus drivers, need to be aware of the movements and actions of people who may be preoccupied with their activities or simply distracted by the spectacle. (Backing a ladder truck into the city manager may sound appealing at budget time but is not a good idea.) Visitors who are not well versed in fireground activities should always be escorted.

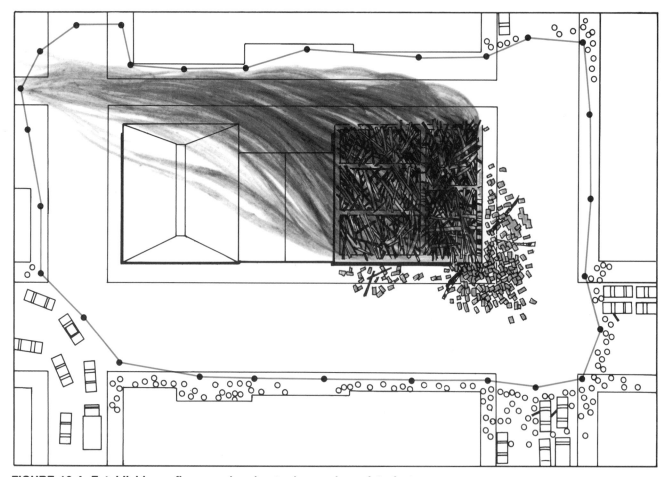

FIGURE 10.4: Establishing a fireground perimeter is a major safety factor.

REHABILITATION

One of the standard characteristics of firefighters is that when asked if they feel okay, they will invariably answer "I'm okay," even if it takes all the strength of their last breath. The FGC and Sector Officers must learn to appreciate this fierce perserverance but must also provide for the welfare of those who refuse to give up until they belong in intensive care. Firefighting is tough and demanding work, and the system must provide for crews to be relieved, rested, or rotated as necessary. This means that the FGC must provide for a *tactical reserve* of personnel to replace them. Sector Officers must routinely monitor the condition of their assigned personnel and advise the FGC when they need to be relieved. The system should provide for full crews to go to a *Rehabilitation (REHAB) Sector,* outside the fireground perimeter, to seek rest, nourishment, comfort, and medical evaluation. At REHAB they should also have SCBAs serviced and refilled. Once they have been through REHAB, crews may become available for reassignment to firefighting action, held in reserve, or released from the scene. If they are too beaten up to rehabilitate, they may be sent to the hospital or given more time to recover their strength.

The system needs to provide for some human factors which reflect simple facts about fireground activities. Regardless of all organizational considerations, safety is concerned with firefighters as individuals and real live human beings. These factors include:

The system needs to provide for some human factors which reflect simple facts about fireground activities. Regardless of all organizational considerations, safety is concerned with firefighters as individuals and real live human beings. These factors include:

- *Fatigue*—When energy levels go down, no amount of effort can achieve normal results. Fatigued firefighters are accidents waiting to happen. They need to be cycled through REHAB and evaluated by the EMTs or paramedics.
- *Fluids and Food*—After a sensible period of labor, dehydration and hunger become urgent considerations.
- *Temperature*—Harsh weather (hot, cold, wet, or windy) quickly affects the condition of firefighters, and the system must protect personnel from long work periods with regular rotation and canteen services.
- *Stress*—The mental condition of personnel must be considered after stressful experiences such as major disasters and life loss situations, particularly the injury or death of other firefighters. These events produce attitude changes, inability to pay attention, lack of control, and errors that produce injuries.

DEPARTMENTAL SAFETY RULES

While the FGC is responsible for managing fireground safety, everyone on the fireground owes the system a strong commitment to safety. This relationship establishes a balance and a sense of shared responsibility. Only the players themselves can actually fulfill the requirements which regulate their own behavior. This depends on their own acceptance of responsibility for operating within the system. The players can break the rules faster than the FGC can catch them, every time.

The system depends on mutual acceptance of certain operating parameters. The basic rules must include:

1. Fire companies operate within certain status categories:
   - In quarters
   - Responding
   - Staged
   - Assigned to an activity
   - Operating within a sector
   - Returning to quarters.

   Regular status does *not* include wandering around, freelancing (pick your favorite activity), hiding out, splitting-up (divided crew), or whimpering and snivelling about the FGC/Fire Chief/Staff.

2. Individual firefighters are operating within the system when they:
   - Wear their protective clothing
   - Use their SCBA
   - Operate with their company or as assigned by their Company Officer
   - Operate tools and equipment carefully
   - Follow SOPs.

They are outside the system when they choose their own plan over the FGC's plan, play daredevil, or compete instead of cooperate.

3. Full protective clothing must include:
   - Helmet with eye protection
   - Hood (or overlapping earflaps and collar)
   - Protective coat
   - Protective pants
   - Boots
   - Gloves
   - SCBA
   - Personal alarm device
   - Friendly teddy bear (for security).

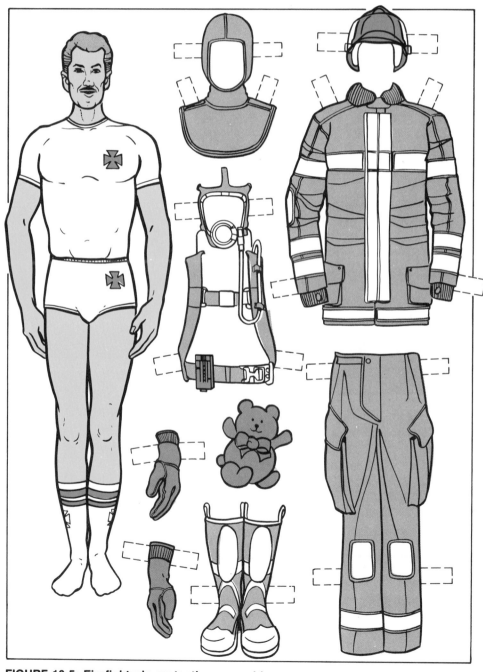

FIGURE 10.5: Firefighter's protective ensemble.

4. Breathing apparatus shall be worn with the face piece in place, using tank air, by all personnel when operating:
   - In a contaminated atmosphere
   - In an oxygen-deficient atmosphere
   - Where both contamination and oxygen deficiency are suspected
   - In an unventilated, confined space
   - Above an involved fire area
   - In an area subject to explosion or sudden contamination.

5. Breathing apparatus shall be worn and ready for use by all personnel operating:
   - Above ground
   - Below ground
   - In an area where the atmosphere may become contaminated

The basic decision on using SCBA is based on three absolute rules:
1. Nobody is ever allowed to breathe smoke
2. Use SCBA until the atmosphere is confirmed to be safe
3. If in doubt—use it.

VEHICLE SAFETY RULES

One of the most dangerous aspects of fireground action revolves around the process of delivering personnel and apparatus to the scene. No matter how serious the emergency may be, nothing positive is contributed by over-enthusiastic participants who injure themselves, destroy apparatus, and wreak havoc on the general population in their zeal to get to the scene.

Before leaving the station, the Company Officer is responsible for verifying that certain requirements have been met, including:

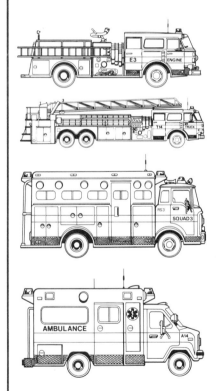

1. The driver is alert and aware of the destination. Drivers should be card-carrying graduates of an official driver training program and checked out on the particular vehicle before they even consider emergency response.
2. The apparatus bay door is open completely, and all crew members are in place and safely belted or strapped to the vehicle. If the driver is in such a hurry that he can't wait for the crew, ask him to explain how effectively he can perform on the scene without them.

Modern fire apparatus provides seats with belts for everyone who should be riding. If older apparatus still requires crew members to ride tailboards, they must be secured by straps, belts, or other positive devices which keep them from falling off. Beware of crew members who shun safety straps and reserve the right to "bail-out" in a tough situation—they are expressing a value judgment on the driver.

When the officer is satisfied that the apparatus is ready to respond, the driver is authorized to proceed to the scene of the incident. Most states' motor vehicle codes give emergency vehicles the legal right of way over other traffic and exempt them from certain restrictions. These provisions are conditional, however, and are in effect only when the vehicle is involved in a valid emergency response, complies with legal requirements for warning devices, and is operated with an appropriate regard for safety.

The driver training program must stress that it is the primary responsibility of the driver to deliver the apparatus and personnel to the scene of the incident, *safely*. Time is of the essence in responding to many emergencies, but high velocity heroics are absolutely unacceptable. The duty of other drivers to yield the right of way is directly related to their ability to perceive, and react appropriately to the emergency vehicle.

The driver of the emergency vehicle must make allowances for whatever anyone else on the road is doing (or not doing) and constantly avoid situations where safe passage does not depend on another driver altering a normal, predictable set of actions. The apparatus driver has to watch out for everyone else on the road, making sure that they are aware of the emergency vehicle and are actually yielding the right of way, before taking advantage of it. This also means that the apparatus driver must avoid taking other drivers by surprise, scaring them into irrational panic actions. Emergency vehicles must be operated in a manner which makes them highly visible and their actions predictable, while allowing for the inappropriate actions of others.

Most fire apparatus is large and heavy, with predictable influence on acceleration, braking, and maneuvering capabilities. We have developed the mechanical engineering capability to power even the largest fire apparatus efficiently to make up in top speed what we are lacking in other areas. When laws exempt fire apparatus from posted speed limits, the burden falls on the driver and the fire department to determine reasonable and prudent speeds for prevailing conditions. Some basic rules must be in effect.

1. On the open road (straight, wide, dry, smooth, good visibility, no other traffic), apparatus must not exceed a set maximum speed. A *limit of 10 m.p.h. above the posted limit.*

2. When normal road conditions are compromised by traffic, weather, disrepair, or any other negative conditions, *the posted speed limit becomes the absolute maximum and actual speed must be regulated by current conditions.* If a vehicle cannot make a turn faster than 25 m.p.h., a legal limit of 35 m.p.h. is only interesting. The driver must keep the vehicle under control at all times.

3. Approaching major intersections, even with the right of way, the apparatus driver must be prepared to stop and never exceed the posted limit. If other apparatus are converging, major intersections can be critical meeting points.

4. Approaching negative right of way situations (red lights, stop signs, railroad tracks), the driver must slow the apparatus to a maximum of 15 m.p.h. and be prepared for an immediate stop. A full stop should be required before crossing in front of any oncoming vehicles and the driver must account for every lane of potential cross traffic. Look out for daydreaming drivers who come from behind a pack of stopped traffic in the open lane. Never try to race the other guy through the intersections (especially trains).

5. Arriving on the fireground, the driver must slow his apparatus to pedestrian speed and maneuver with due regard to all of the other activity which may be going on. Running into or over each other is not acceptable.

PHYSICAL FITNESS

The fire department must have *standard levels of physical fitness* for personnel involved in emergency activities. Obviously, this does not mean that only Olympic medal winners can be firefighters; however, everyone must be able to perform at a level that makes the difference between being an asset or a liability.

Physical fitness programs should promote strength, flexibility, and endurance in a reasonable balance. This may be achieved through a variety of activities which straddle the line between recreation and flagellation. Running, weight-lifting, and racquetball are generally considered to be positive approaches to physical fitness—shooting pool, belly-butting, and combat basketball are not.

MEDICAL CONTROL

An *annual* medical examination, based on set standards and performed by a physician familiar with the occupational health and safety factors of firefighting, should be mandatory for every firefighter. Particular attention must be directed toward the cardiovascular and respiratory systems before an individual is authorized to participate in fireground activities or use SCBA.

Personnel recovering from injuries or major illnesses should be re-evaluated by the fire department physician before they are approved to participate in fireground activities. This is not a rule based on punishment—it is meant to protect the players from their own enthusiasm to get back into action.

SUMMARY

To insure firefighter safety, the FGC must be concerned with firefighter attitude, Command attitude, safety programs, safety training, assignments of Safety Officers, use of self-contained breathing apparatus (SCBA) and protective clothing, and equipment use and maintenance.

The FGC can re-enforce safety training through the use of operations critiques and reviews. On the fireground, the FGC must make all risks become controlled risks, justified to save lives and valued property.

Early size-up, adequate crews and support, and proper equipment and training are essential for primary searches. The search teams must have a specific plan to search a specific area. The search for victims must never be done at the expense of the search teams.

A safety program must be designed to prevent death and injury. It should detect and correct dangerous problems, be concerned with firefighter fitness and health, and stress proper equipment usage and readiness. The program must develop a standard set of safety procedures.

Safety training must include vehicle safety, use and maintenance of equipment, fire station safety, and physical conditioning.

Vehicle safety on or near the fireground can be greatly improved if the FGC uses staging with an assigned officer.

A Safety Officer should be assigned for complex fires, structural hazard fires, hazardous materials incidents, and any incident where the FGC needs a Safety Sector.

The FGC must demand that SCBA be worn for fires above or below ground level, operations in contaminated atmospheres, and all situations where atmosphere contamination is likely.

It is a safety function of the FGC to demand that full protective clothing be worn at all times by those operating on the fireground.

Equipment and apparatus failures must be noted. It is necessary to follow-up on replacement and repair.

The fireground is the area inside of an imaginary line that has been determined by safety considerations. To adequately define the fireground, the FGC must consider foreseeable hazards for a particular incident.

When setting the perimeter, the FGC must take into account the location of the fire related to the center of the street, the clear area available, and the areas of potential explosion, collapse, and smoke drift.

All personnel entering the fireground must wear protective clothing, have their crews intact, and be assigned to a sector.

The operating area of crews is one of the most important safety considerations. Great caution must be exercised when companies are placed above the fire floor, where fire can move in behind them, when combining interior and exterior attacks, where access and egress are limited, and when fighting below-ground fires.

For operations above and below ground level, two separate, far apart escape routes are required.

Where appropriate, an aggressive interior, offensive attack may eliminate most safety problems before they occur. Firefighter safety is the critical factor in deciding on offensive or defensive modes.

Command must continually evaluate the tenability of a structure for interior attack. The 20 minute rule must be applied for most structures.

Defensive operations should be far away from the fire as is practical, utilizing available barriers.

Safety is highly dependent on controlling personnel. The FGC should limit personnel on the fireground to those people assigned to a necessary task. Hazardous situations require that the absolute minimum number of personnel be inside the fire perimeter.

Establishing sectors controls the position and function of the firefighters. The Sector Officer must be able to account for the location and welfare of every worker in every crew assigned to his sector. When necessary, a personnel identification system must be used.

Firefighter safety is improved through the use of communications to coordinate activities. Certain actions, such as the use of ladder pipe streams while ground crews are operating on the fireground, demand efficient communications.

The FGC uses the regular sector system to order an evacuation. The emergency traffic announcement is used for fast evacuation during critical times. During evacuation, the Sector Officer should move crews to safety and then verify the safe exit of sector personnel.

COMMAND DEVELOPMENT

Firefighter safety must be the number one priority on the fireground. To be an effective Fireground Commander, you must know how to safely manage personnel through the use of sectors. You must determine as early as possible if a Safety Sector is needed.

One of your first tasks on the fireground is to establish a fireground perimeter. You must know the safety factors to apply for each particular situation. Practice is easy, since the building does not have to be on fire to be a useful example.

Ordering a primary search creates anxiety for many new FGCs. You must practice your ability to do size-ups and to notice foreseeable hazards. Ordering a primary search requires you to be able to set objectives for the search and rescue teams, limiting their activities to a specific area. Simulations help to develop the skills you will need. Study how decisions were made for ordering the primary search for every fire you attend.

Whenever you consider how the FGC interacts with Sector and Company Officers, look for the ways in which such interactions promote safety. Make certain that standard safety procedures be a part of all Command SOPs.

When you practice critiques and reviews, attempt to apply each procedure to the effort of improving safety for the firefighters. Consider their safety both on and off the fireground.

The following report card is provided so that you can evaluate your FGC knowledge and skills in classroom exercises, simulations and on the fire scene.

Fireground Commander Report Card

Subject: Safety

Did the Fireground Commander:

- ☐ Correctly establish a fireground perimeter?
- ☐ Consider crew safety when ordering:
    - ☐ a primary search?
    - ☐ offensive attacks?
    - ☐ defensive attacks?
    - ☐ salvage overhaul?
- ☐ Utilize staging for safety?
- ☐ Establish sectors?
- ☐ Establish a Safety Sector and assign a Safety Officer when needed?
- ☐ Utilize communications to promote safety?
- ☐ Limit personnel on the fireground.
- ☐ Use standard safety procedures as part of the SOPs used for an incident?
- ☐ Use critiques and reviews to improve firefighter safety?

TIMELESS TACTICAL TRUTHS

For the past 20 years (or so), I have jotted down notes on 3 X 5 cards relating to firefighting and fireground command. Most of these comments have occurred to me as I have watched, listened, pondered, and generally been involved in firefighting operations. I have captured these observations in the odd places and at the strange times when such thoughts seem to strike. Recording these thoughts through the years has resulted in a giant stack of cards. These short shots are fireground items that don't exactly fit into any place in the text but describe some fireground reality. They are short, very simple, and somewhat sarcastic (my weakness).

I realize that presenting a batch of one liners at the end of a book is somewhat unusual, but I am concluding with these comments so I can finally get rid of the blasted cards. I will probably start a new collection because the habit of instinctively taking notes is hard to break. If any tactical one liners occur to you, please send them along or save them until we meet down the line.

Thanks for looking the book over. I hope it helps you on the fireground.

- Better to be too big, instead of too small.
- Move quick—young conditions are easier to control than old ones.
- The very worst fireground plan is no plan (the next worse is two plans).
- The only safe fireground assumption is to assume the worst.
- If you have lots of ideas, you need lots of companies.
- A little force in the beginning can eliminate the need for lots of force at the end.
- It is better to get out five minutes too soon than five seconds too late.
- Very little on the fireground falls up.
- Respect defensive conditions—the buildings God didn't want to burn are sprinklered.
- Don't ever let your inclination to gamble outdistance your fear.
- Safe firefighters are smart firefighters.
- Vomiting firefighters are ugly firefighters.
- Fires give the test just ahead of the lesson.
- The only thing that will impress the fire is well-placed force (force = water).
- Never confuse repeat fires for routine fires; the same basic deadly elements are present at *every* fire—there are *no* routine fires.
- Don't stand too close to the guys who are always bandaged up.
- If you panic, be certain to run in the correct direction.

- Safety prevents meetings.
- 1st rule of exits: If you pay to get in, you have paid to get out.
- You can't save anyone when you are a victim.
- Always avoid hanging around guys they call "Burn 'em Down Brewster," "Charcoal Charley," or "Parking-Lot Pete."
- Gravity will always culminate at the bottom.
- 3rd law of radiant heat: When your warning lights begin to melt, it's a sign that you parked too close.
- Effective command is made up of equal parts of passion and patience—the trick is the where and when of each.
- Unless the walls are falling, the FGC shouldn't yell or run— neither reflects cleverness or composure.
- The capable FGC always approaches his troops with high expectations and kindness.
- You can fool the spectators, but you can't fool the players.
- All fires go out eventually.
- Trust safety, not luck (luck makes you DUMB).
- Something is wrong if you keep inheriting bad situations.
- Playing catch-up on the fireground is an old female dog.
- Firefighting is the very smartest form of manual labor—respect the task.
- Don't assume anything is separated—the only perfect fire walls are in heaven.
- When you're having problems, take on a partner to share them.
- Don't spend all your chips—always have a tactical reserve.
- Remember courage is only fear that has said its prayers.
- Complicated fire operations are generally screwed up fire operations.
- Losing your temper generally represents the incipient stage of rectal-cranial inversion.
- Good procedures are so simple you don't need to write them down to remember them or use a dictionary to understand them.
- When someone screws up, yell at them—they'll love it.
- Keep working on the basics—most of us are not advanced enough to make advanced mistakes.
- The fire always plays for keeps and is unforgiving and democratic—it will never kick you when you're up.
- Be careful of the guys who close their eyes when they open their mouths.
- Don't add resources to a non-plan.
- Be careful who you give water to.
- Educational times on the fireground are not always fun times.
- Consistent fireground reality: if you're gonna order, you gotta pay the check.
- Be careful of kamikaze pilots who have gone on 65 missions.
- Burning up all your exposures at once is tacky.

- The fewer parts of the plan, the fewer things can get screwed up.
- The number of faults in a fire operation is in direct proportion to the number of viewers; the intelligence of the viewers is in direct proportion to how late they arrive.
- You gotta have a plan before you can revise it.
- Retreat festivals are far superior to funeral festivals.
- Experience and education are like oregano—they must be mixed with a lot of other stuff to be good.
- It's difficult to get a little excited.
- Be careful where you put water.
- The treatment for screwed-up situations: education, training, reflection, and getting to do it again.
- Be careful of people who attach status to knowing things you don't.
- Combining strategic modes (offensive/defensive) is like ordering artillery on yourself.
- The essence of fire fighting is that the fire and the FGC can't live in the same space—one has to leave.
- Beware of the Chief who says, "Don't do anything until I get there."
- Never is a long time.
- The role of the effective FGC is to direct and support the troops.
- Avoid the folks who say regular safety procedures take too long during difficult times (when you really need them).
- Standard management cycle: procedures—training—execution—critiquing—revising and back around.
- Firefighters who are tough enough to eat nails and spit pumpers will generally become extremely fragile when mistakes occur.
- Manage procedures—lead people.
- There isn't any middle ground in fire fighting—you're either fighting or you're not fighting.
- If you think training is expensive, check out the cost of ignorance.
- Forget the baloney about "holding the fire"—you either put it out or it burns past you.
- There ain't no fair fights on the fireground.
- What burns, never returns.
- Avoid situations that are so exciting you don't survive.
- Some days on the fireground the best it gets is so-so.
- Extinguishing the fire in most cases solves the majority of problems.
- Smart FGCs practice "chocolate chip management"—when the troops do good, give 'em a cookie.
- Effective analysis must always be mixed with water to put the fire out.
- When the pipe goes up, the building comes down.
- Don't ever develop a plan that is so smart you can't explain it to the people who have to carry it out.
- It's hard to legislate love.
- Procedures don't have feelings.

- Water above a fire basically irritates Mother Nature—she meant heat to rise.
- You can't fall through from down below.
- When you lose your head, the next thing is your ass.
- If a building burns, don't take it personally (you didn't make the world combustible).
- If looking at a fire makes you crazy, don't look at it.
- Don't hang around the folks who substitute their ears for a standard safety procedure—basic safety rule: don't ever burn *any* of you.
- If you can't control yourself, you can't control anything else.
- Surprises are nice on your birthday.
- If you live with a bad situation long enough, you wear it.
- Effective communications = 1 part talking and 10 parts listening (beware of the guy whose hearing is affected by promotion).
- Hanging around daredevils is painful—remember, a hero is nothing but a sandwich.
- It's hard to generate a big fire in a small building.
- Take the process seriously—not yourself.
- The things that lead up to accidents happen slowly—the accident happens fast.
- Everything on the fireground is "too" something.
- Always take care of the people who are trying to make you look good (make it as easy as possible for them to do so).
- Hope for the best—plan for the worst.
- Most big screwed-up situations start with one small out-of-balance step in the wrong direction—be careful of confusion snowballs that start rolling downhill.
- The FGC must beware of the altitude affecting the reasoning power of the roofers working above the fire: lots of big topside mistakes occur on the roof—most of them seem to involve water.
- Be careful of anyone who thinks they have nomex skin.
- There are no credit cards on the fireground—you've got to pay for everything you do at the time you do it.
- The more seniority a screw up gets, the harder it is to fix—this applies to both the firefighting operations and (unfortunately) firefighters.
- The FGC has to always go after the right piece of information at the right time; if you ask if everything is okay, it always is.
- If you aren't dressed to play, stay in the bleachers and off the field.
- Roofs are really pretty dumb—they shed ladder pipe water just like it was rain.
- Considering what was going on at the time a decision was made will many times effectively refocus 20-20 hindsight.
- Repeat, ongoing business can become tactical anesthetic that produces a fireground sucker punch: like 20 bells and smell (nothing) calls in the same building setting you up for the 21st that turns out to be a ripper.

- There aren't any "time outs" on the fireground.
- The next tragedy will take the pressure off the last tragedy.
- The FGC must always initiate and move toward correct action. He must also be prepared and capable of stopping incorrect, unsafe action—he absolutely cannot live with a bad situation.
- A FGC who will not disagree with a decision or countermand (and change) an order, should stay home and watch the fire on TV.
- They generally don't call the fire department because someone did something smart.
- Be careful of shutting down and unhooking anything that is set up and operating okay.
- The FGC should be the first person who thinks the fire is burning and the last to believe it is out.
- Every fire situation contains a discreet number of decisions—they can be made either by the FGC or by the fire.
- Do not think you are communicating just because you are talking.
- Most of the time on the fireground, the first five minutes are worth the next five hours.
- When someone screws up, ask the standard question, "Who taught him how to do it?"
- The longer you wait to make a decision, the fewer options you will have.
- The FGC must be careful of what he says in difficult situations— offhanded, dumb command comments are like aluminum beer cans—they last forever in the environment.
- Basic fire frequency axiom: the farther you are from the last fire, the closer you are to the next one.
- The FGC must always be able to separate what is a hope from what is a plan.
- There is a big difference between an air tank SOP and a face piece SOP.
- There is no necessary connection between the amount of hose in the street and the amount of water that goes on the fire.
- The smart guys on the fireground can tell what is going to happen— any dope can tell what has (already) happened.
- Don't ever trust smoke—it can hide what's really going on, spread the fire, burn, blow up, and really ruin your life, sometimes all at once.
- The FGC must always have a "string" on his troops—be careful of any situation where you can't get the insiders out quickly and account for them.
- The only thing that won't burn are the bricks.
- If a fire is an emergency to the fire department, who would you call?
- If you violate, compensate.
- The more routine decisions the FGC makes before the fire, the more time he will have to make critical decisions during the fire.
- The most important fire is the next one.
- Life isn't made up of only good stops (life is not perfect).

- A brick falling directly on your head will often ruin your day.
- Don't change a rule by breaking it.

- If you must burn a building, do it with class!

SAFETY TRUTHS

- Think.
- Drive defensively.
- Drive slower rather than faster.
- If you can't see, stop.
- Don't run for a moving rig.
- Always wear your seat belt.
- Wear full turnouts and SCBA.
- Attack with a sensible level of aggression.
- Always work under sector command/no free lancing.
- Don't *ever* breathe smoke.
- Keep your crew intact.
- Maintain a communications link to command.
- Evaluate the hazard—know the risk you're taking.
- Follow standard fireground procedures; know and be part of the plan.
- Vent early and vent often.
- Use a line that's big enough and long enough.
- Do not ever go beyond your air supply.
- Always have an escape route (hose line/life line).
- Provide lights for the work area.
- If it's heavy, get help.
- Always watch your fireground position.
- Look and listen for signs of collapse.
- Rehab fatigued companies—assist stressed companies.
- Pay attention all the time.
- Everybody looks out for everybody.

BIBLIOGRAPHY

Readers interested in additional books on firefighting and fire command may find the following classic titles to be of special interest. Several of these titles are now out of print but should be available in libraries.

Brannigan, Francis L. *Building Construction for the Fire Service,* 2nd ed. National Fire Protection Association, Quincy, Massachusetts, 1982.

Bryan, John L. and Picard, Raymond C., eds., *Managing Fire Services.* International City Management Association, Washington, D.C. 1979.

Casey, James F. *The Fire Chief's Handbook,* 4th ed. Technical Publishing Co., New York, 1978.

Clark, William E. *Fire Fighting: Principles and Practices.* Dun-Donnelley Publishing Corp., New York, 1974.

Coleman, Ronny J. *Management of Fire Service Operations.* Duxbury Press, North Scituate, Massachusetts, 1978.

Davis, Larry W. *Rural Firefighting Operations,* 2 vol. International Society of Fire Service Instructors, Ashland, Massachusetts, 1985.

Fried, Emanuel. *Fireground Tactics.* H.M. Ginn Corp., Chicago, 1972.

Kimball, Warren Y. *Fire Attack One.* National Fire Protection Association, Quincy, Massachusetts, 1966.

_____ . *Fire Attack Two.* National Fire Protection Association, Quincy, Massachusetts, 1968.

Layman, Lloyd. *Fire Fighting Tactics.* National Fire Protection Association, Quincy, Massachusetts, 1953.

McAniff, Edward P. *Strategic Concepts in Fire Fighting,* McAniff Associates, Bayside, New York, 1974.

Page, James O. *Effective Company Command.* National Fire Protection Association, Quincy, Massachusetts, 1974.

Walsh, Charles V. and Marks, Leonard G. *Firefighting Strategy and Leadership,* 2nd ed. Gregg Division, McGraw-Hill, New York, 1977.

GLOSSARY

Adequate Resource Stage—occurs when sufficient units have arrived to cover both key forward and back up jobs.

Alarm—fire department communications center.

"All Clear"—the primary search has been completed.

Apparatus Placement—the location and utilization of fire apparatus based on the categories of responding, staged, operating, parked, and returning to quarters.

Area Arrangement—the streets, buildings, potential exposures, and access obstacles.

Attack Plan—a systematic plan developed by evaluating conditions, developing tactical approaches, identifying tactical needs, identifying available resources, and making assignments. It is based on the three tactical priorities.

Attack Plan Variables—the location/position of attack, size of attack, support functions, and the time of attack.

Attack Plan Worksheet—a standard set of review items, serving as a flowchart for completing the evaluation of the attack plan and organizing the required revisions.

Backdraft—an explosion that may result from the sudden introduction of oxygen into a space-restricted fire.

Benchmarks—the objectives of each tactical priority; the primary search is completed, the fire is controlled, and loss is stopped.

Command—fireground radio designation for the FGC. Refers to the person, the functions, and the location of command.

Command Mode—one of the three commitments of the first-arriving Company Officer. Strong, direct command is required from the outset.

Command Modes—three possible commitments to be made by the first-arriving Company Officer. These include the nothing showing mode, fast attack mode, and the command mode.

Command Post—the standard position for the FGC, usually stationary, inside the command vehicle or fire apparatus.

Command Vehicle—vehicle used for the transportation of command officers and for storage of prefire plans and other significant fire data. Generally used as an initial command post.

Company Officer—works under the command of a Sector Officer. Commands at the task level.

Decentralized Command—see Sectors.

Defensive Strategy—an exterior attack, with related support, designed to stop the forward progress of the fire and then provide fire control.

Deficient Resource Stage—usually during the initial stages of active fires where there are more jobs to do than companies to do them.

Direction of Attack—the position attack companies assume in approaching the fire. This direction involves actual fire stream placement, and the location of support functions.

Elevated Master Streams—defensive fire streams provided by ladder pipes, platforms, buckets, and booms.

"Emergency Traffic"—a priority message to be immediately broadcast throughout the fireground.

Engine Company—basic unit of fire attack consisting of apparatus and personnel trained and equipped to provide water supply, hose lines, location and removal of endangered occupants, and treatment of the injured when necessary.

Engine Company Officer—fire department officer or firefighter responsible for apparatus placement and operation and tactical deployment and safety of engine company personnel.

Fatigue Stage—when workers begin to tire and their safety and welfare must be protected. The FGC utilizes staged reserve companies.

Fire Attack (classical)—the ideal division of work where engine companies advance hose lines while ladder companies clear the way and open up just ahead of them.

Fire Control Forces—fireground personnel who are involved in gaining entry and access, providing support activities, and control of the fire.

Fire Control Sector—responsible for an effective fire attack, stopping extension, and extinguishing the fire.

Fire Control System Model—defines the necessary management activities and the relationships between them. SOPs are its foundation.

Fire Extent and Location—how much and what part of the building is involved.

Fireground—defined by an imaginary line (fireground perimeter) which encloses the space where the fire situation creates potential hazard to fire personnel.

Fireground Commander (FGC)—the person who assumes overall command and control of personnel and apparatus at the emergency incident scene. He assumes the role of commander and manager, operating at the strategic level.

Fireground Factors—a list of the basic items that the FGC must consider when evaluating tactical situations. These factors are building, fire, occupancy, life-hazard, arrangement, resources, action, and specific considerations.

Fire Line—marks the area reserved for the fire department to operate without having to deal with spectators, traffic, and other problems.

Fire Stream—water applied directly to the fire to control the fire.

Fire Stream Factors—fire stream type, size, placement, timing, and supply.

Flashover—gases trapped against the ceiling ignite, quickly involving the entire interior space.

Forcible Entry—activities required when firefighters encounter barriers that keep them from the fire area.

Fully Involved—Immediate entry and search activities are impossible and victim survival is improbable. The affect of the fire is such that an "all clear" will not follow.

Functional Sectors—assigned to perform specialized tasks or activities which do not necessarily coincide with geographic sectors.

Geographic Sectors—responsible for all general firefighting activities in an assigned area.

Initial Report—a short radio transmission to provide a description of conditions and the confirmation and designation of command.

Interior Access—normal means of access, including stairs, hallways, interior public areas, etc.

Interior Sector—responsible for operations within the fire building.

Key Hydrants—those hydrants located closest to the fire in safe positions.

Key Tactical Positions—those spots on the fireground that place apparatus in the best position to work to capacity.

Ladder Company—basic unit of fire attack consisting of apparatus and personnel trained and equipped to provide location, protection and removal of fire victims, provision of forcible entry and gaining access, ventilation, checking for fire extension, control of utilities, operation of elevated master streams, salvage, and overhaul.

Level I Staging—initial arriving attack teams (e.g., an engine and a ladder or an engine and a rescue squad) go directly to the scene taking standard positions, assume command and begin operations. The remaining units stage about one block from the scene, until ordered into action by the FGC.

Level II Staging—used for large, complex, or lengthy operations. Additional units are staged together in a specific location under the command of a Staging Officer.

Life Safety Decision—the FGC's decision as to whether to remove the victims from the fire, the fire from the victims, or some combination of the two.

Marginal Mode—fireground period occuring at the end of the offensive stage and the beginning of the defensive stage. This can be a dangerous period for interior operation.

Mushrooming—a build-up of the products of combustion that have charged the top of the fire area and then spread down laterally.

"Nothing Showing"—a "very minor fire" that allows for an interior search until it can be reported "all clear." Usually, occupants will not have to be removed.

Offensive Strategy—an interior attack, with related support, designed to quickly bring the fire under control.

Police Liaison Sector—established by the FGC to coordinate crowd and occupant control.

Postfire Critique—a process that examines the effectiveness of fireground operations. The critique is designed to reinforce positive performance and focus on lessons learned.

Prefire Plan—written analysis of the fire problems of a particular building in terms of size, hazards, and built-in protection.

Primary Fire Damage—the damage produced by the basic products of combustion.

Primary Goal—the safety of all the participants.

Primary Search—a rapid search of all involved and exposed areas affected by the fire that can be safely entered. Its purpose is to verify the removal and/or safety of all occupants. Occupant status can be verified on every offensive operation, whether or not actual fire is involved.

Property Conservation—the third tactical priority designed to reduce primary and secondary damage and to allow for scene preservation.

Reconnaissance—information not visually available to the FGC at the command post. It is usually acquired by assigning personnel to specific problems and receiving their reports.

Rescue Company—apparatus and personnel trained and equipped to provide ladder company or other specific functions on the fireground as determined by the FGC.

Rehabilitation Sector—an area outside the fireground perimeter where crews can go for rest, nourishment, comfort, and medical evaluation.

Rescue Mode—the stage of activities on the fireground until the primary search is completed.

Rescue Order—a structure for the initiation of rescue activities and the evaluation of resources needed, based on actual and potential rescue needs.

Rescue Sector—responsible for location, protection, and removal of fire victims.

"Right Way" Direction of Attack—a smooth attack through the unburned portions of the building, pushing the fire upward and out through ventilation openings.

Safety Factors—the critical elements of a fireground safety program that include command attitude, firefighter attitude, SOPs, Sector Officers, safety training, health and fitness, SCBA and protective clothing, equipment and apparatus, and risk management.

Safety Officer—a specialist who provides expertise and individual attention to supplement the role and responsibility of the FGC for fireground safety. He should work as a sector within the fireground organization.

Secondary Fire Damage—the damage caused by rescue, support, and fire control operations.

Secondary Means of Rescue—using elevated platforms, aerial ladders, fire escapes, ground ladders, helicopters, etc., when normal means of exit are not possible or practical.

Secondary Search—a complete, thorough search of the interior fire area after completing fire control, ventilation, and other required support activities.

Sectors—a smaller, more manageable unit of fireground command delegated by the FGC to provide management and command for specific functions or geographical areas of the fireground (see Geographical Sectors and Functional Sectors.)

Sector Officer—assigned by the FGC to manage specific geographical areas of the incident scene or specific fireground functions. He operates at the tactical level.

Size-up—the initial phase of the situation evaluation.

"Smoke Showing"—the conditions exist where it is possible to extend rescue and fire control simultaneously to gain entry and control interior access. The rescue mode is in effect until the primary search is completed and an "all clear" is transmitted.

Standby Stage—when assignments are basically stable and the FGC has more companies than jobs, a tactical reserve can be created.

Staging—the management of committed and uncommitted apparatus to provide orderly deployment. See Level I Staging and Level II Staging.

Staging Officer—advises the FGC of equipment and resources available, assigns specific companies to the FGC's requests, and assists these companies in responding to their assignments.

Standard Operating Procedures (SOPs)—a set of organizational directives that establish a standard course of action on the fireground to increase the effectiveness of the firefighting team. They are to be written, official, applied to all situations, enforced, and integrated into the management model.

Strategy—the management of the offensive/defensive decision by the FGC. This critical decision regulates operational control, establishes objectives, sets priorities and allocates resources.

Stream Effective Reach—the maximum distance at which a hand line will be effective.

Support Activities—the quick development of resources needed to support the attack, e.g., ventilation, forcible entry, and provision of access.

Tactical Level—operated by Sector Officers who have been assigned to specific areas and tasks by the FGC in order to meet operational objectives.

Tactical Priorities—the ordered sequence of rescue, fire control, and property conservation.

Tactical Sequence—1. rescue, 2. fire control, 3. property conservation.

Tactical Worksheet—a systematic-approach worksheet that is designed to allow the FGC to have a standard way to write and record all important fire activities.

Task Level—operated by fire companies, involving the evolution-oriented functions needed to produce task-level outcomes. The Company Officers report directly to the Sector Officers in their assigned area.

Ventilation Factors—the need to ventilate, timing, type, and the correct organization of support units.

Ventilation Types—horizontal, vertical, mechanical, hydraulic (water fog).

Ventilation Profile—can roof operations be conducted?

Victims—those members of the public who are inside, trying to escape from, or outside the fire building who are directly affected by the fire or the losses caused by the fire.

Visual Observations—direct FGC observations of area arrangement, fire building detail, fire conditions, resource status, and the effects of firefighting action.

Water Supply—providing sufficient water (agent) for firefighting as soon as possible.

"Working Fire"—see "Smoke Showing."

INDEX

– NOTES –

- NOTES -